Illustrator:
Howard Chaney

Editorial Project Manager:
Paul Gardner

Editor:
Kathy Humrichouse

Editor in Chief:
Sharon Coan, M.S. Ed.

Creative Director:
Elayne Roberts

Associate Designer:
Denise Bauer

Cover Artist:
Susan Williams

Product Manager:
Phil Garcia

Imaging:
James Edward Grace

Acknowledgements:

HyperStudio® is a registered trademark of Roger Wagner Publishing, Inc.

ClarisWorks software and screenshots are © 1991-95 Claris Corporation. All Rights Reserved. *ClarisWorks* is a registered trademark of Claris Corporation in the U.S. and other countries.

Astound® is a registered trademark of Astound Inc.

Apple the Apple Logo and Macintosh are trademarks of Apple Computer, Inc., registered in the United States and other countries.

Publishers:
Rachelle Cracchiolo, M.S. Ed.
Mary Dupuy Smith, M.S. Ed.

INTEGRATING TECHNOLOGY INTO THE MATH CURRICULUM

INTERMEDIATE

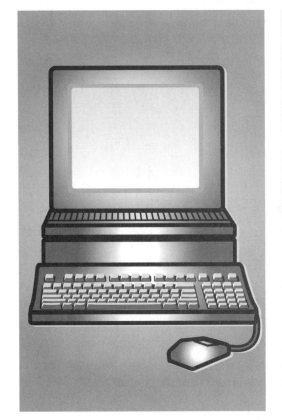

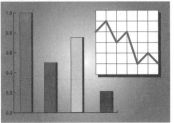

Author:

Dennis Soares

Teacher Created Materials, Inc.
6421 Industry Way
Westminster, CA 92683
www.teachercreated.com

ISBN-1-57690-425-3

©1998 Teacher Created Materials, Inc.
1999, Reprint
Made in U.S.A.

The classroom teacher may reproduce copies of materials in this book for classroom use only. The reproduction of any part for an entire school or school system is strictly prohibited. No part of this publication may be transmitted, stored, or recorded in any form without written permission from the publisher.

TABLE OF CONTENTS

Introduction 5
Suggested Software and Hardware 6
Classroom Management Ideas
- One Computer Classroom 7
- Multiple Computer Classroom 8
- The Computer Lab 9

Schedule
- One Computer Classroom 10
- Multiple Computer Classroom 11

Planning 12
Assessment 13

THE UNITS

Working With Money
- It All Adds Up 14
- The Toy Store 16

To Add or Subtract
- Number Fairy Tale 19
- Slide Show Quizzer 21
- Number Line Math 23

Everyday Estimations
- My M&M's 26
- A Random Number Generator 28

Multiplication And The Microprocessor
- Math In A Flash 31
- Multiplication Bingo 34

Tech Time
- Time Marches On 36

Dive Into Division
- Show Me How 41

Finding Averages The Easy Way
- Not Your Average Board Game 44

TABLE OF CONTENTS

Fractions Can Be Fun
- The Fraction Machine 48
- Fractions In A Box 50
- Fractions In A Circle 52
- All Mixed Up 54

Decimals And Disk Drives Go Together
- Traveling Decimals-Tenths 57
- Traveling Decimals-Hundredths 59
- Self-Examination 61

Probability- Nothing More Than Chance
- Take A Chance 62
- Manipulative Madness 64

Measuring Matters
- Working With Rulers 67
- Measuring My Stuff 69
- Finding The Perimeter 70
- Finding The Area 72

Going Geometric
- Crossword Geometry 74
- Hyper Geometry 76

Working With Charts And Graphs
- Charting Children 80

Working With Ratio And Percentage
- Ratio 84
- Percentage 83

Math Journals And Dairies
- Keeping Track 89
- Talking Journals 90

Math Puzzles
- Mystery Squares 91
- The Secret Code 93

TABLE OF CONTENTS

Map Skills
- Multimedia Research 95
- Mapping The World 97
- Mapping My City 99

Finding Math In Sports
- Super Sports League 102
- Watch Randy Run 105

As Simple As Giving Directions
- Duplicate Me 109
- My Very Own Robot 111
- The Way to School 112
- Pet Expert 113

As Simple As Following Directions
- Really Reading 114
- Check Out These Checkers 116

Activities Using Logic
- If-Then Statements 120

Student Created Computer Programs
- Hyperstudio Adding Machine 124
- An Astounding Program 130

For Teachers And Parents
- Project Checklists For Teachers 136
- Scheduling 139
- Planning 140
- Project Answers (for pages 114-116) 141
- Software Bibliography 142
- Other Resources 143
- Bibliography 144

INTRODUCTION

As we inch closer toward the 21st century, technology continues to play a greater and more significant role in our lives. Each year scores of educational institutions across this nation, and throughout the world, connect to the Internet. The increased demands placed on teachers, particularly at the K-12 level, inhibit extensive study and planning for the use of technology in the classroom. The problem is compounded when teachers are not comfortable using the technology they currently have. *Integrating Technology into the Math Curriculum—Intermediate Level* will provide teachers, technology coordinators, paraprofessionals, students, and parents with well thought out solutions to using technology in a meaningful way.

The opening section of this book will suggest ways to use this material. A listing of software that is appropriate for the lessons will also be found here. Classroom management is critical in getting maximum results from the technology that is available. This issue is addressed in the opening section of this book. Ideas for one-computer classrooms, multi-computer classrooms, and computer labs are presented.

One of the things that concerns many people in education today is how to accurately measure the effectiveness of using technology. The evaluation process is essential if learning is to take place. Students need to do more than just complete a worksheet using technology. They also need to evaluate and analyze the results. They need to ask themselves the question, "How can I use what I learned from this lesson in real life?" To this end, each unit in this book will have an evaluation section and follow-up activities.

Included after the introductory section, are thematic units that require the use of a computer. Units are comprised of 2-3 activities and include the grade level, time frame, suggested software, materials needed, assessment/evaluation activities, and Internet addresses of a math site(s) with more information on each topic. Teacher Created Materials attempts to offset this ongoing problem by posting changes of URL's on our Web site. Check our home page at *www.teachercreated.com* for updates on this book. The steps for doing each activity will be broken down as follows: Before the Computer, On the Computer, Assessment and Follow-up Activities. *Integrating Technology into the Math Curriculum— Intermediate Level* is an excellent resource for ways to use technology in conjunction with the elementary math curriculum.

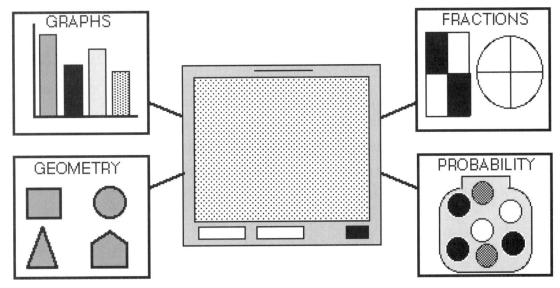

Introduction

SUGGESTED SOFTWARE AND HARDWARE

Before using this book in a classroom setting, take an inventory of the software and hardware available. Here is a list of some of the software titles that would work well with the activities found throughout *Integrating Technology into the Math Curriculum*. Keep in mind that there are also numerous "shareware" and "freeware" titles that would also work.

Hardware:

Macintosh- LC, Performa, Quadra, and Power PC

IBM (or clone)-386 or better using Windows

Video capture card

Digital camera

Color printer

Scanner

Modem for Internet access

Software:

Type To Learn- Sunburst Corporation

The Writing Center- The Learning Company

Kid Pix- Broderbund Software

Kid Pix Studio- Broderbund Software

Hyperstudio- Roger Wagner

Kid Desk- Edmark

At Ease- At Ease

HyperCard- Apple

Astound- Astound Incorporated

Claris Works- Claris Corporation

Claris Draw- Claris Corporation

Microsoft Works- Microsoft Corporation

File Maker- Claris Corporation

Calculator- Apple

Graphic Converter 2.06 (c)- Thorsten Lemke

Grolier's Multimedia Encyclopedia- Groliers

Compton's Encyclopedia- Compton's

Lotus Smart Suite- IBM Corporation

Microsoft Paint- Microsoft Corporation

Future Basic- Future Basic

CLASSROOM MANAGEMENT IDEAS

The One-Computer Classroom:

Managing a classroom of 20-33 children can be a challenge as any teacher knows. When a computer or two is thrown into the mix, the challenge becomes even greater. The issue boils down to how to use the computer without it interfering with the daily lesson plans and the instruction the students need to receive. Here are some ideas that may help:

1. Pair students up. Put a strong computer user with a weaker one. It may take some time and experimentation to determine who these students are.

2. Schedule students in pairs to work on the computer for 30-45 minute intervals. Depending upon the number of students, it may be possible to schedule several computer sessions for each group during a single week.

3. When setting up a classroom computer schedule, try to avoid those times when students will be silently reading. This is usually the first subject taught in the morning.

4. Allow time in the computer schedule for the computer to be used as a "free time" activity or incentive.

5. Make sure there are a few "experts" in class who will be the first people those at the computer will go to if they get stuck. The "experts" should be trained in how to restart the computer, find necessary programs, and help interpret the instructions of the lesson. They will also make the teacher's life a lot easier.

6. Most programs have a "sound off" and "sound on" function. This feature can be accessed from within the computer's control panel. When the computer is set up for the day (usually in the morning), turn the sound off. If sound is needed, inexpensive headphones will do the trick.

7. If the teacher wants to model a lesson before it is assigned, connect the computer to a television or projection screen. On an IBM (or clone) simple converter boxes can be purchased for as little as $79.00. For the Macintosh converter boxes are in the $100.00 range. These boxes will allow the computer to be connected to a TV/VCR. The lesson could also be recorded for classroom showing later. LCD projection units start at $1,300.00 and go well over $7,000. These allow what is on the computer monitor to be displayed on a large overhead screen.

8. With some experimentation, there is a way to schedule computer time in the classroom. Do not forget to use parent volunteers and aides when they are available. Look at the schedules on the following pages for more ideas.

Management

CLASSROOM MANAGEMENT IDEAS (Cont.)

Multiple-Computer Classroom:

If there are several computers in the classroom, the teacher will have greater flexibility when making up the computer schedule. In some situations, there may be a LAN (Local Area Network) or a WAN (Wide Area Network) already set up at the school site or computer lab. When working with Macintosh computers, it would be well worthwhile to investigate "Local Talk" and the networking capabilities already built into the machine. Some suggestions for working with multiple computers in the classroom:

1. Organize the class into groups or teams. Groups could pick their own team names.
2. Make sure that the groups or teams contain students of mixed abilities, ethnicity and gender.
3. When making up the weekly schedule, be consistent. For example, "blue group" should go on the computers at the same time on Monday, Wednesday, and Friday.
4. Once again keep the computer time in the 30–45 minute range for students at this age level. Impress upon them the need to accomplish the goals/objectives for that block of time. They should be working collaboratively. Students who are absent *may* be able to make up their time in one of the free time blocks that have been reserved.
5. Avoid scheduling computer time during reading. Reading is the most fundamental subject, and all students should participate. It is often the first subject in the morning.
6. Remember to turn the sound off on the computer when it is not needed, or provide the students with headphones.
7. If "centers" are used in the classroom, the computers could become one of the centers.
8. If students are going to be saving their work, be very clear with them *how* and *where* it should be saved. Saving and loading work should be a lesson itself. When using a management tool such as "Kid Desk" or "At Ease", it is possible to set up, in advance, save and load locations for work created by the students.

Networking Ideas for Multiple-Computer Classrooms:

When working with several Macintosh computers in the classroom that are not networked together, here is what will be needed:

1. A phone-net device for each computer and printer that will be part of the network. These sell from $9-30.00, depending on the brand. Essentially, one of these devices is plugged into the printer port on each computer. A regular phone line is strung from one computer to the next in a series.
2. From the Apple Menu select the Chooser Item. Click on the button to turn on Apple Talk. The computer may need to be restarted at this point to make sure "Apple Talk" is activated. Next go back to the Apple Menu and select Sharing Setup from the control panel. Start file sharing now. Be sure to give the network a name.
3. Go back to the Chooser again and click on the icon for Apple Share. The names given to the computers at the Sharing Setup window, will now appear.
4. It may also be helpful to investigate Users and Groups and File Sharing Monitor found in the control panel. The main advantage of networking is the ability to share hard drives and student work. It does take a little practice, however.

Management

CLASSROOM MANAGEMENT IDEAS (Cont.)

The Computer Lab:

Some elementary schools have chosen to organize the computers into a computer lab. This would be a room that has been designated for the housing and operation of computers and computer related technology. This approach can be very effective, if resources are limited, in order to get the maximum use out of an investment. However, a word of caution is necessary. Too often computer labs have no real vision and become little more than drill and practice centers. A computer lab assistant, who is in charge of maintaining the lab and providing instruction, could be useful. In order for a computer lab to be successful in integrating technology across the curriculum, a great deal of thought and planning must go into how it is going to be used.

Philosophy and Purpose:

In simple but clear terms, define what the vision is for the computer lab. Form a technology team, if one does not already exist, to articulate the goals and objectives for each grade level in the core curriculum. By setting grade-level appropriate goals and objectives that correlate to the core curriculum, educators will ensure that everyone is "on the same page." Naturally, when the results of the technology program are evaluated, goals and objectives can be adjusted. The idea is to consistently apply the school's philosophy and purpose to all students. As students move through the elementary grades, teachers will know what computer skills they have already learned.

Software:

When selecting software, there are only two categories to address. The first category is instructional software. This kind of software usually takes a concept and reinforces it or extends it. Simulations and problem-solving activities can be found in some instructional software, while other titles may be little more that drill and practice. The second category is productivity software. There seems to be an increasing trend toward this type of software especially at the elementary level. The reason for this is obvious. Productivity software allows the user to be the author or creator of something. Most students enjoy this because it takes them from spectator to active participant. As software has become better and more sophisticated, there seems to be an endless variety of choices when it comes to using productivity software. For the purposes of this book, much of the software recommended does fall under the category of productivity software.

Equipment:

The kind of equipment used is less important than the understanding of how to use that equipment. So often schools may have decent equipment but lack teachers on site who really know how to use the equipment in an effective manner. If the equipment is old and outdated, make the best of it while at the same time setting goals for future needs. Visit schools that have a technology program that is really making an impact. There are numerous ways to raise money for technology including bake sales, book club sales, collaboration with supermarkets, and numerous grant opportunities. In addition, many prison systems are involved in refurbishing computers and distributing these to schools. These computers are often very useful in the hands of someone who understands what to do with them. Lastly, get to know those people on site, at other sites, at the district level, and at the county office, who are knowledgeable about technology. They will provide a wealth of information and guidance in the decision making process.

Management

SCHEDULING - ONE COMPUTER CLASS

The form below can be used to make out a computer schedule for a class. Remember it is a good idea to pair up students when only one computer is available. Another option is to group students into teams. This works particularly well if the students are already sitting in groups of 4-6. Since keyboarding is essential to nearly everything one does with a computer, it is important to have students work with a typing program at the beginning of each computer session.

Mixed Ability Groups

Tigers	*Warriors*	*Fireflies*	*Mustangs*
Shelly	Sylvia	John	Cindy
Jose	Bounthavy	Jennifer	Outhay
Robert	Rosemary	Theresa	Patti
Ione	Charles	Victor	Sam
Dennis	Matthew	James	Andrew
Roger	Cynthia	Darin	Tegra

Weekly Class Computer Schedule

Time	Monday	Tuesday	Wednesday	Thursday	Friday

SCHEDULING - MULTIPLE COMPUTER CLASSROOMS

In a multiple-computer classroom, students will be able to spend a lot more time on the computer. Assign students to work at a computer station and insist that they only work at that computer when it is their turn. This makes it easier to retrieve documents and check the status of projects. A management system such as *Kid Desk* or *At Ease,* is recommended for teachers. These types of programs limit student access to the hard drive and manage where files are saved. They will save a great deal of time and "suffering" if they are used correctly.

If the school is wired for the Internet, use a computer-naming convention that is easy to remember and will also identify the computer. For example, if room 5A had five computers, the computers could be named 5A-1, 5A-2, 5A-3, 5A-4, and 5A-5. Be sure to change the name of the hard drive. Also, go into "Sharing Setup" found in the "Control Panel", and make sure that the hard drive name is used here as well before file sharing is started. Remember if the school is hooked up to the Internet, students can eventually send and receive mail. Thus, someone sending mail to 5A-3 would know that the e-mail address was something like *5A-3@Stockton.K12.ca.us.* A well thought out naming convention when using multiple computers, will make things a lot easier later on.

Weekly Class Computer Schedule

Group 1
_____
_____
_____
_____
_____
_____

Group 2
_____
_____
_____
_____
_____
_____

Group 3
_____
_____
_____
_____
_____
_____

Group 4
_____
_____
_____
_____
_____
_____

Group 5
_____
_____
_____
_____
_____
_____

Group 6
_____
_____
_____
_____
_____
_____

Time	Monday	Tuesday	Wednesday	Thursday	Friday

Insert time and group number.

Management

PLANNING

It's a good idea to go through the activities in this book before asking students to do them. If the instructions are unclear to the teacher, the students will probably be confused. How this book will be incorporated into the curriculum will change depending on the classroom setup and the number of available computers. Teachers may also want to collaborate with each other when using this material.

Suggestions For Using The Lesson Plans In This Book

These lesson plan ideas are intended for a variety of situations that might be encountered at an elementary school site. Some planning in advance is required to do many of these activities, and the students must have a minimum understanding of using computers and software. Simple exposure to some of the suggested software and what can be done with it, will be most helpful.

Here are some ideas that can be used when incorporating these lesson plans into the curriculum:

 Decide on the amount of time that each pair, group, or the entire class is going to receive on the computer(s). In the classroom, use some kind of timer to keep everyone on task. Students will also need to decide who will go first, second, and so on. At the end of the project, each student should indicate in writing how they contributed, in addition to completing the assessment portion of the project.

 When starting a new lesson, briefly introduce it to the whole class so that everyone has a feel for the objective of the lesson. Make sure that students are aware of the time line and that they make use of the little box found on each lesson plan stating what kind of software would be appropriate for that lesson plan. At this point, allow students to work independently, only going to the "classroom experts" when they are really stuck.

 Schedule student-teacher technology conferences. This is a time for teachers to meet with students one-on-one to share their projects. Since many of these lesson plans ask for a printout of the results, a special place in the classroom can be created where the results can be posted.

 Use the "Project Checklist for Teachers" planning sheet found on page 136. With this sheet, teachers can track all of the students in the room and either mark projects complete/incomplete or give letter grades. Keep in mind that these lesson plans do not necessarily follow any kind of sequential learning order. It is perfectly acceptable to jump around to cover these topics during the regular math curriculum.

Management

ASSESSMENT

When a teacher is ready to start integrating technology into the math curriculum, he or she must assess the students' achievement. When working with technology projects, do not be afraid to use alternative assessment methods. Students cannot solely be evaluated with some kind of test where only one correct answer might exist. Multimedia projects may look vastly different from other each but still complete the objectives of the lesson. Think of assessment as a way of monitoring a student's progress by collecting information and data about a project.

The kinds of assessment tools that will be found in this book, will focus more on the process rather than the product. There are also checklists and rubrics to monitor student performance. Because these units encompass many skills and may require anywhere from one day to several days to complete, evaluation of student performance must be flexible.

Try breaking a project down into smaller parts and then determine what should be evaluated. For example, if the student is working on a unit in geometry and chooses to use *Hyperstudio* as the software tool, an assessment of the project might look like this.

EXAMPLE OF A HYPERSTUDIO PROJECT

Prior Planning:

Did the student plan in advance and clearly understand the purpose and goal of the lesson or project?

Use of Technology Tools:

Was the student able to use both software and hardware in an effective manner?

Content:

Was the content provided by the student (the finished product) the proper content for this lesson or project?

Creativity:

Does the finished product show signs of creativity or imagination in the way in which the instructions were interpreted?

Technical Skills:

What was the overall quality of the production? Were the requirements for grammar and syntax met?

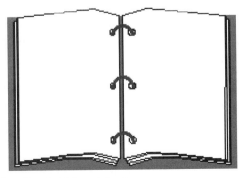
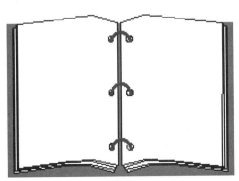

Working With Money

IT ALL ADDS UP

Using the computer, students will combine coins to come up with as many possible ways to make $1.00 as they can find.

Grade Level: three to five

Duration: 40-60 minutes on the computer

Materials: samples of U.S. coins, Color Code and Combinations Charts (Pg.15)

Software: any drawing program such as *Kid Pix*, *Hyperstudio*, *Claris* (must be able to move objects)

Internet Links: http://pages.prodigy.com/kidsmoney/kids.htm
http://www.usmint.gov/

Procedure:

Before the Computer:

- Start this activity by providing students, either in groups or individually, with samples of all current U.S. coins. Attempt to get all students to differentiate between a half dollar, quarter, dime, nickel, and penny. Also discuss the relative values that these coins have (one nickel equals five pennies, one dime equals two nickels).

On the Computer:

- Each student will create 2 half dollars, 4 quarters, 10 dimes, 10 nickels, and 10 pennies. Students must keep the proportions between the coins the same (i.e. a dime is not bigger than a nickel. Using the Copy and Paste functions found in nearly all drawing programs will make this an easy task.

- Students may want to make each coin a different color and create a color code to differentiate between coins.

- Once all of the coins have been created, the task is to combine them in as many ways as possible so that the **sum** equals $1.00. The student must move the coins around on the screen to come up with a combination that totals $1.00. Record this combination on the sheet that is found on the next page, and then look for other combinations. After all possible combinations have been found, be sure to **save their work**.

Challenge:

- What combination requires the greatest number of coins to make a sum equal to $1.00? With a partner, students can move several different coins to the center of the screen and try to tell what the value is of these coins. Find out what all of the coins add up to!

Assessment

- Students should print out the computer screen and answer these questions in complete sentences, either at the bottom of the paper or on the back. Turn in this chart.

#2425 Integrating Technology Into the Math Curriculum © Teacher Created Materials, Inc.

Working With Money

IT ALL ADDS UP (Cont.)

1. What was the most difficult part of this activity?
2. Is there another way to do this activity?
3. Make up a game to play with a partner using what was created on the computer screen.

Color Code Chart: Use this chart to make a color code of the coins.

COIN COLOR	Half-Dollar	Quarter	Dime	Nickel	Penny

Fill in this chart with combinations that equal $1.00

Attempts	Half-Dollar	Quarter	Dime	Nickel	Penny	Total $
#1	2					$1.00

Working With Money

THE TOY STORE

In this lesson, students will create their own toy store. This store will contain 10 different kinds of toys for sale, each marked with a different selling price. Each student will have exactly $20.00 to spend, and must use as much of his or her money as possible to buy toys. Skills include using estimation, addition, subtraction, and following directions.

Grade Level: three to five
Duration: 60–90 minutes on the computer
Materials: samples of coins and bills, Toy Store Log Sheet (Pg. 17)
Software: any drawing program such as *Kid Pix, Hyperstudio, Claris Draw,* etc., Computer Calculator, CD-ROM clip art.
Internet Links: http://www.childwood.com/

Procedure:

Before the Computer:

- Ask students how many of them like to go shopping for toys (most hands should go up).
- Find out what some of their favorite toys are and how much they cost.
- Review or teach the concept of estimation.
- Have students make a list of toys they would want in their store.

On the Computer:

- Students can use a desktop calculator or the calculator that is on the computer.
- Using a drawing program such as *Kid Pix*, students will create a store window in which there are 10 different kinds of toys for sale. They can use the pre-made stamps in *Kid Pix*, CD-ROM clip art, or draw their own toys. Keep in mind there is a time limit of 90 minutes for this activity.
- They may make multiple copies of their toys. For example, they could create 3 dolls or 2 basketballs The students should decide what toys to price at the following values and record this on the next page. If the drawing program also has the ability to enter text, they should label each toy with the value they have chosen. Remember to use the Copy and Paste functions to speed up the work.
- They will need to price the toys using the following **exact values**.

Toy-1	$.99	Toy-6	$4.59
Toy-2	$1.49	Toy-7	$5.29
Toy-3	$2.17	Toy-8	$6.99
Toy-4	$3.49	Toy-9	$7.99
Toy-5	$3.99	Toy-10	$9.99

- Put together as many toys as possible without going over $20.00 Record this on the Toy Store Log Sheet. Now put together another combination of toys that comes close to $20.00 without going over. Complete the forms on page 17.

Working With Money

THE TOY STORE (Cont.)

Toy Store Log Sheet

MY TOYS

Describe your toys **Toy Price $**

Toy-1 _____
Toy-2 _____
Toy-3 _____
Toy-4 _____
Toy-5 _____
Toy-6 _____
Toy-7 _____
Toy-8 _____
Toy-9 _____
Toy-10 _____

The Toy Store Log Sheet			
Shop	List toys that you bought here	Cost $	Change $

Working With Money

THE TOY STORE *(Cont.)*

Assessment:
Have the students print out the computer screen and answer these questions in complete sentences either at the bottom of the paper or on the back.

1. How did you determine which toy went with which price?
2. How could rounding and estimation help you with this activity?
3. Did you use the calculator or did you figure the amounts by hand?
4. What is the greatest number of toys that you can buy?

Follow-up Activity:
- Set up a token economy in the classroom. This is a system where tokens or play money are used in the classroom as incentives.
- Find out what kinds of things students would like to have for sale in their store. Items could be stickers, pencils, books, erasers, folders, etc.
- Each week (or whatever period of time decided) the store would open for business.
- Different pairs of students would run the store, handling transactions (sales) and giving out change.

Example Screen—"It All Adds Up" **Example Screen- "The Toy Store"**

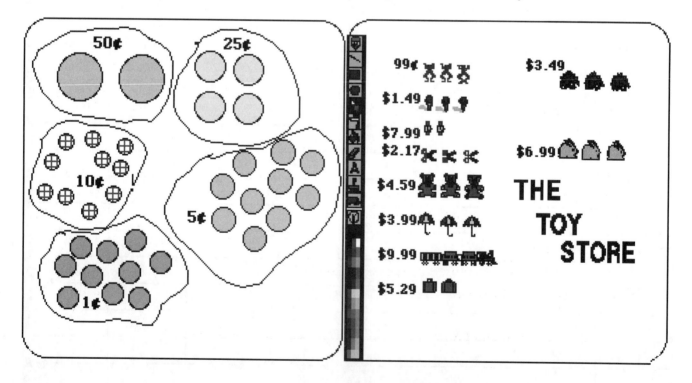

To Add or Subtract

NUMBER FAIRY TALE

Students will be integrating writing into the math curriculum by using the computer to create a short story using numbers as the main characters.

Grade Level: three to five
Duration: 30–60 minutes on the computer
Materials: computer, scanner, digital camera, Number Character Traits Sheet (Pg. 20)
Software: any program such as *The Writing Center, ClarisWorks,* or *Simple Text,* that will allow you to write and save text.
Internet Link: http://juliet.stfx.ca/people/fac/pwang/mathpage/math1.html

Procedure:

Before the Computer:
- Teachers should tell students that they are going to create a fairy tale involving three numbers as the main characters and read them a classic fairy tale as an introduction to this lesson.
- Teachers should ask students to select **three** numbers from 0–100. Have them make one number the **good** number, another the **sneaky** number, and the third the **wise** number.
- Students will make a list of the qualities for each number ahead of time and record the qualities on the sheet provided. Students will also create an illustration using each number as the main shape.

On the Computer:
- Using the computer and a word processor, students will create a short number story (three paragraphs minimum) about the numbers that they have chosen. The story should have a problem and a solution. Students should use the character traits from their list to tell about the number characters.
- If a scanner or digital camera is available, scan or take a digital image of drawings that students made earlier and add them to the story. Students may also use a drawing program to create their illustrations,
- If the writing program has a spell-checker, students should check their work before printing.

Assessment:
Students will assess their own work using the Self Assessment Survey on page 20. In each category they will enter a score, then add up their total and write it on the top of the printed copy of their story.

Follow-up Activities:
- Combine all the Number stories and make a class book.
- Find out which numbers were the most popular and why. Then make a graph of the distribution of numbers chosen.
- Have students do an oral presentation of their story or create a skit about what happens in their Number story.

To Add or Subtract

NUMBER FAIRY TALE *(Cont.)*

NUMBER CHARACTER TRAITS		
Wise _____	Good _____	Sneaky _____
LIST TRAITS	**LIST TRAITS**	**LIST TRAITS**

Self-Assessment Survey: Number Story

Did I follow all directions? (0–25 POINTS)	Does my story make sense when others read it? (0–50 POINTS)	Quality of the artwork: Did I follow all directions? (0–25 POINTS)	Total Points (Not greater than 100 points)

To Add or Subtract

SLIDE SHOW QUIZZER

In this lesson, students will develop a slide show math quizzer that will show their understanding of addition and subtraction and allow them to test other students.

Grade Level: three to five
Duration: 60–90 minutes on the computer
Materials: computer, clip-art, Slide Show Sequence sheet (Pg.22)
Software: software that can generate a slide show such as *ClarisWorks, Hyperstudio, HyperCard, Kid Pix Studio,* or *Astound*
Internet Link: http://www.csun.edu/~vcact00g/math.html

Procedure:

Before the Computer:
- Teachers should review the concepts of addition and subtraction, of up to three digits, and regrouping.
- Students may need some experience using the slide show function found on programs like *Claris Works, Hyperstudio, Powerpoint,* etc. It is essential that they understand how to make a series of screens or cards and set them up to play at timed intervals.
- Students will use the Slide Show Sequence sheet on page 22 to plan how the cards will look. There will be five problem cards and five answer cards. Do not skip this step! Teachers should initial students' papers when they have completed their pre-planning.
- **Students must have at least one of each kind of the following problems:**
    a. Addition problem with single digits (example: 8+7=15).
    b. Subtraction problem with single digits (example: 9-3=6).
    c. Addition or subtraction problem with two digits (example: 22+34=56).
    d. Subtraction problem that requires regrouping (example: 45-19=26).
    e. A problem in which the sum or difference is greater than 100.

On the Computer:
- Each student will be making ten screens or cards from the plans that they drew up in the previous section.
- Students will create the problem screen followed by the answer screen. Remember, there are five kinds of problems they must create.
- Set the time interval between the problem screen and the answer screen for 10-20 seconds. This should give others an opportunity to guess the answer.
- Students should think of ways they can designate groups of ten so that they will not need to draw each and every shape or paste dozens of pieces of clip art.

To Add or Subtract

SLIDE SHOW SEQUENCE

1.

2.

3.

4.

5.

6.

7.

8.

9.

10.

To Add or Subtract

NUMBER LINE MATH

The objective of this activity is to integrate the math skills of addition and subtraction into the Social Studies curriculum.

Grade Level: three to five
Duration: 60–90 minutes on the computer
Materials: computer, clip-art, Time Line Number Chart (Pg. 24)
Software: multimedia Encyclopedia such as *Compton's* or *Grolier's,* drawing program or word processor
Internet Link: http://www.safari.net /~rooneym/

Procedure:

Before the Computer:

- Teachers should review with students what a number line is and how it is used. Tell them that they will be making an historical time line and will then use this information to solve addition and subtraction problems. Show them some examples of number lines.
- If students have not worked with a multimedia CD-ROM before, go over with them how to search for information. They will be looking for specific information and recording the dates that these events happened on a number line about famous individuals in history.

On the Computer:

- The first step is to construct a number line using either a drawing program or a word processor. Students will go to the Page Setup menu and choose to have the page in Landscape mode rather than Portrait. This will allow the page to be wide across and narrow on the right and left edge. The teacher or class expert can help if students have a difficult time with this. Students should use as much of the paper as possible when drawing the number line.
- Next, label the time periods on their number line. This number line will only extend from 1000 AD to 2000 AD. It is important to get the spacing between the numbers even. In a drawing program, use the rulers or auto grid function. In a word processing program, use the Tab key.
- Make the farthest point to the left the year 1000, and then evenly space the following years moving toward the right (1100, 1200, 1300, etc.). When working with a drawing program, students can also put a dot on the line near each number (see the next page for an example).
- Using a multimedia CD-ROM, students need to look up the birthdates of the following famous people. When they have found a birthdate, students should place the name of the person in the correct location on the line in parentheses with their birthdate, Joe Smith (1887).
- After students have completed their number lines, they can do some addition and subtraction problems with the dates they have found. They should fill in the Time Line Number Chart using the information from their number line and print a copy of their work.

To Add or Subtract

NUMBER LINE MATH *(Cont.)*

Find the following birthdates and create your time lines similar to the sample below.

1. Johann Sebastian Bach
2. Salvador Dali
3. George Washington
4. Christopher Columbus
5. Sally Ride
6. Sir Frances Drake
7. Pope Adrian V1

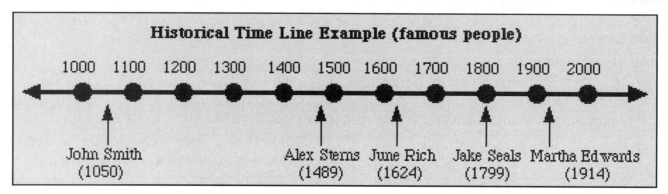

Directions: Create 9 addition and subtraction problems by placing the birthdates from your number line in the boxes below. Find the solution for each problem. Follow the example.

Time Line Number Chart

Year 1	(+ OR -)	Year 2	=
1624	-	-1050	574

To Add or Subtract

NUMBER LINE MATH *(Cont.)*

Assessment:

Accuracy is very important when doing this activity. Students need to find the correct birthdates for the individuals listed. When performing the operations, the sums and differences also need to be accurate. The number line needs to be in correct proportions and the years equally spaced. This can be difficult for students this age to do. When names and birthdates are placed on the time line, they need to be in the proper location. To assist in grading this activity, a rubric such as the one below may be used.

Rubric For Assessing Number Line Math

Rate each of the sections of the activity on a 0-4 scale.
(0-not done, 1-poor, 2-satisfactory, 3-good, 4-exceptional)

Activity	Score
1. The student drew a number line with the correct years labeled and proper spacing. The line was done in landscape mode on the page.	
2. All seven of the listed names were located on the number line in the proper position.	
3. The worksheet on page 24 was completely filled out and the sums or differences were correct.	
4. The number line was printed and turned in with page 24.	
Record the total score here. -------------------------→	

Follow-up Activities:

- This activity lends itself to research projects. Students could select any one of the names from the list on page 24 and do a research paper on this individual using The Writing Center or some other word processor. Some things to include in the paper would be:
    a. place of birth and year
    b. important accomplishments or contributions to humanity
    c. family information
    d. death-how and when

Everyday Estimations

MY M&M'S

In this lesson, students will be provided with the opportunity to estimate and make predictions based upon those estimations.

Grade Level: three to five
Duration: 60–90 minutes on the computer
Materials: computer, M&M's plain (one package per child), Student Prediction Sheet (below)
Software: *ClarisWorks, MSWorks,* or any program that has drawing and graphing ability (charts may also be drawn on the computer)
Internet Link: http://www.eduplace.com/math/brain/

Procedure:
Before the Computer:

- When using a program such as *ClarisWorks* or *MSWorks,* review with the students how to create a chart or import a chart into a document. If they are working with a word processing program that also has drawing tools, show students how they could create a chart or bar graph. The directions will describe a bar graph.
- Review estimation. Why do we use estimation and how can this skill save us time or make work easier?
- Students need to fill out the Student Prediction Sheet found on this page before they start the activity on the computer. They will be predicting the number of colors and number of M&M's of each color in their bags. This is done **before** the bag is opened.
- Have students open the candy bags *carefully,* and sort the candy into groups according to colors. Record the actual number of candies for each color beneath the estimate or prediction (see sample on page 27).

Student Prediction Sheet

	Name	Estimation	Actual	Rank
Color-1				
Color-2				
Color-3				
Color-4				
Color-5				
Color-6				
Color-7				
Color-8				

Everyday Estimations

MY M&M'S *(Cont.)*

On the Computer:

- Now they are ready to start graphing the results. The following chart was created in *ClairsWorks*, but any graphing program will work.

- There are numerous kinds of charts and graphs that students can make from this data. The graph such as the one shown below, is a bar graph. Check the index of the program students are using to see how to create this type of graph.

- It might look something like this.

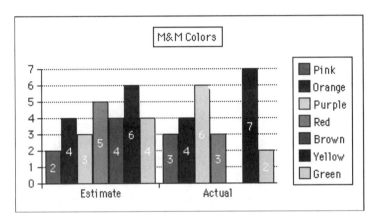

	A	B	C
1		Estimate	Actual
2	Pink	2	3
3	Orange	4	4
4	Purple	3	6
5	Red	5	3
6	Brown	4	0
7	Yellow	6	7
8	Green	4	2

Assessment:

The assessment for this activity will consist of two parts. Students will print their graphs and then respond to the following questions, either on the back of the paper or using the computer.

1. How did the actual results match your guesses? Were any correct?

2. Did you guess the number of colors used in M&M's correctly? Add up the total number of M&M's, and compare this number to the sum of your guesses. Were these two numbers the same or different?

3. Compare your results to other student teams in the classroom. Can you find a color that seems to be most often used? What is this color? Do you think there is any reason that M&M's contain this color more than any other color?

4. What other candies could you have used instead of M&M's to do this activity? Would sunflower seeds have worked? Why or why not?

Follow-up Activities:

- Substitute other kinds of candies and do the same activity.
- Collect the data from all the students in the class, and make a class graph showing class results.
- Have a class debate about why a particular color M&M is the best or favorite among students.
- Put all the individual graphs into a slide show using ClarisWorks, MSWorks, Power Point, Astound, or some other presentation tool.

Everyday Estimations

A RANDOM NUMBER GENERATOR

This activity will allow students to use computer programming techniques to create a random number generator. They will be using this generator to round numbers to the nearest 10's, 100's, 1,000's, etc.

Grade Level: three to five

Duration: 60-90 minutes on the computer

Materials: computer, Rounding Chart and My Action Plan (Pg. 30),

Software: *Hyperstudio 3.0, HyperCard 2.3, Visual Basic,* or any program that allows for the creation of random number.

Internet Link: http://yn.la.ca.us/cec/cecmath/math-elem.html

Procedure:

Before the Computer:

- Students need to have a fair understanding of how to round off numbers. The random number generator can be modified for the lower grades so that fewer digits are displayed.

- This lesson will be using *Hyperstudio 3.0* by Roger Wagner. *Exploring HyperLogo,* which is also published by Roger Wagner, is not normally included with the purchase of *Hyperstudio,* but it includes many scripting commands that can make working with *Hyperstudio* much more powerful. Even if students are not familiar with *Hyperstudio,* this short lesson will be easy to follow.

On the Computer:

- The first step in this activity is for students to create their very own random number generator. While there may appear to be many steps, students will be working with two kinds of *Hyperstudio* objects, besides graphics.

- Students begin by creating a new stack and picking a background color for the first card.

- From the Objects menu select Add A Text Object. Make this object long and narrow (see screen shot for more detail). Give this text object the name Number. Select the Read Only and Draw Frame check boxes. Do not select the other two choices.

- Click on the Style button and choose a font size of 72. Also pick a color. There will be an option here to have the text aligned center, left, or right. Choose centered.

- Now add a button from the Objects menu and name it Create a Number. Students can use an icon if they wish.

Everyday Estimations

A RANDOM NUMBER GENERATOR (Cont.)

On the Computer: *(Cont.)*

- Students should now be at the Things To Do and Places To Go window. The stack will only have one card, so choose Next Card in Places To Go. In Things To Do, choose a sound. Now select automatic timer and set it for 20 seconds.
- Finally, choose the Use HyperLogo box. Students may have never used this screen before, but they will see a blank window. Type the following text **exactly** as it is printed here: **setfieldtext [] "Number Random(100000)** This simple code tells HyperLogo to generate a number between 1 and 99,000 each time the Create Number button is pressed. Students will have 20 seconds to write this number down and then round it off on their worksheet.
- The [] symbols are located right next to the P key on the keyboard.
- If students have followed the previous steps carefully, the *Hyperstudio* card should look something like the one on page 28.
- Students should decide who will go first, then click on the Create button. The other student will record this number on the form on page 30, and he will have 20 seconds in which to round the number.
- After twenty seconds a new random number will appear. This time the students switch and the other student records the number and then rounds it in the same manner.
- Be careful to round the numbers off to the **correct place** as requested on the worksheet.

Assessment:

Students will check each other's worksheets to see if rounding has been correctly done. Teachers can use the random number generator to do a whole class activity or assessment to find out students' understanding of rounding. The time interval could be changed to make rounding the numbers more difficult.

Follow-up Activities:

- Have students make a list, in their journals, of all the ways that rounding numbers could be used in everyday life.
- There are numerous ways that a random number generator can be used. With the information that students were given on how to make a random number generator, put together a card that will help you decide the following:

1. What to play for P.E. (hint-1 could be kick ball, 2-could be basketball, 3- could be dodge ball, etc.).
2. Students could use a random number generator to play a board game. They can create a simple board game using *Hyperstudio* or some other drawing program, and then use the random number generator to play the game.

- Be sure to save the Random Number Generator so that it can used for future activities.

Everyday Estimations

A RANDOM NUMBER GENERATOR *(Cont.)*

Rounding Chart

Highest Place		Nearest Hundred		Nearest Thousand		Nearest Ten Thousand	
1,232	1,000	42357	42,400	176,899	177,000	44,222	40,000

My Action Plan:
I plan to do one of the challenges or one that I have thought of on my own.

Name: _____ Date: _____

_____
_____
_____
_____
_____
_____
_____

Multiplication and the Microprocessor

MATH IN A FLASH

Students will be creating a multiplication facts slide show or program using a variety of software and hardware tools

Grade Level: three to five
Duration: 90-120 minutes on the computer
Materials: computer, clip art, digital camera or scanner, Multiplication Planning Table (Pg. 33)
Software: *Hyperstudio 3.0, HyperCard 2.3, Visual Basic, ClarisWorks, Kid Pix, Astound,* or other program that creates slide shows.
Internet Link: http://forum.swarthmore.edu/dr.math/dr-math.html

Procedure:

Before the Computer:
- Students need to make a list of what facts they wish to put into their slide show.
- If the drawings are going to be created by hand and either scanned or captured with a digital camera, they will need to be done at this time.

On the Computer:

- Depending upon the selected program, students have many choices to make with regards to the design and operation of the Math in a Flash slide show. The following example will use *Hyperstudio,* although a number of other programs will work just as well.
- Students should create blank cards for each fact.
- Here is an example of a *Hyperstudio* screen. By using invisible buttons that have an automatic timer to move from card to card, students can string together a number of these types of screens as a slide show, each with a different multiplication problem.
- Each student might also want to add a sound clip, stating the problem and the answer for each card. This could also be attached to the invisible button.
- Graphics can be any PICT or GIF formatted picture. Students can also draw graphics in *Hyperstudio* or add them from the clip art folder.

© Teacher Created Materials, Inc.

Multiplication and the Microprocessor

MATH IN A FLASH *(Cont.)*

On the Computer:

- If students are working with programs like *MSWorks, ClarisWorks,* or *Kid Pix*, they could put together screens like this one which was done with *ClarisWorks*. Notice that a different method of grouping was used here.

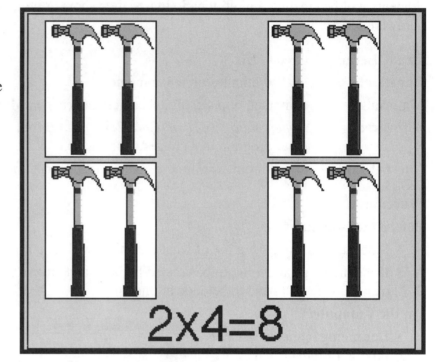

- The number of cards or screens they put together should be decided before going on the computer.

- To quickly turn this into a training program for muliplication, leave out the answer when making the *Hyperstudio* stack. Have a timed button say the answer after a set interval of time. This will require that students use the Play A Sound function.

- Students could also turn this into a multiple choice testing stack by creating three buttons on each card. Give one button the correct answer and the other two incorrect answers. Assign a Boing sound to the incorrect answer buttons and the word Correct as the sound to play when the correct answer button is pressed.

Assessment:

In assessing this activity, teachers should look for three things. First, did the student follow the directions and are the multiplication examples correct? Second, is the project neat and well organized showing that thought and consideration were given to the placement of text and graphics? The third area involves the quantity of screens generated. Obviously there is a difference between a student who created 20 screens and one who only created three.

Follow-up Activities:

- This project could be a long-term one that students would work on throughout the school year. The goal would be to get all of the times tables through the 10's in a slide show format. This kind of undertaking would require some thought. They do not want to make each succeeding screen so predictable that it would be very easy to know what the answer is. In order to prepare for the sequence of screens, they need to do some pre-planning. Here's how!

    1. Make a list of all the multiples from 0x0 to 10x10.

    2. Using the form on page 33, pick a problem at random and enter it into a location on the grid. Cross this multiple off the list that they made. They may have noticed that many multiples are really the same, the numbers are just inverted.

Multiplication and the Microprocessor

MATH IN A FLASH *(Cont.)*

Mulitplication Planning Table

0x0										

Mulitplication Planning Table

0x0										

Multiplication and the Microprocessor

MULTIPLICATION BINGO

In this lesson, students will use the random number generator from page 28, and will play multiplication bingo.

Grade Level: three to five
Duration: 30-60 minutes on the computer
Materials: computer, Random Number Generator (see pg. 28)
Software: *Hyperstudio 3.0, HyperCard 2.3, Visual Basic,* or any other software that allows for the creation of random numbers.
Internet Link: America Online - *Student Center*

Procedure:

Before the Computer:

- Teachers may want to go over the rules for the game of bingo. In this case, students will be looking for the multiplication fact that corresponds to the number generated.
- Reproduce the bingo cards provided on page 35 or have students make their own.

On the Computer: (This is a duplication of the instructions on page 35.)

- Students begin by creating a new stack and picking a background color for the first card.
- From the Objects menu select Add A Text Object. Make this object long and narrow. Give this text object the name, Number. Select the Read Only and Draw Frame check boxes. Do not select the other two choices.
- Click on the Style button and choose a font size of 72. Also pick a color. There will be an option here to have the text aligned center, left, or right. Choose centered.
- Now add a button from the Objects menu and name it Create a Number. Students can use an icon if they wish.
- Students should now be at the Things To Do and Places To Go window. The stack will only have one card, so choose Next Card in Places To Go. In Things To Do, choose a sound. Now select automatic timer and set it for 20 seconds.
- Finally, choose the Use HyperLogo box. Students may have never used this screen before, but they will see a blank window. Type in the following text exactly as it is printed here: text **exactly** as it is printed here: **setfieldtext [] "Number Random(145)**
- This simple code tells *HyperLogo* to generate a number between 1 and 144 each time the Create Number button is pressed. They can change the 144 to any number that they wish, but remember not all numbers will be in the times tables (prime numbers).
- Students are now ready to play with a partner, team, or entire class. If they are using the pre-made bingo cards, they will find problems already printed, such as 3x7=. When the number 21 comes up, they can put an X on this problem since this is the solution. Students take turns clicking on the Create Number button until someone wins.

Multiplication and the Microprocessor

MULTIPLICATION BINGO *(Cont.)*

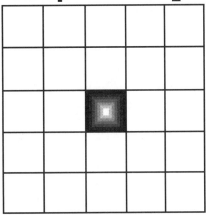

Assessment:

Students would be assessed according to who won the game of bingo. It is important to keep track of the numbers played so that cards can be checked later to see if the problem crossed out really does match a number that was created by the random number generator. By changing the number 145 in the random number generator script, it is possible to work with much larger numbers. The chances of having an answer on the bingo card, will decrease the greater the random number.

Follow-up activities:

How could this same activity be used to work with addition, subtraction, and division problems? How would students design the random number generator? What changes would they make? How would they design the bingo cards?

Students are probably familiar with the game of tic-tac-toe. Could they use the random number generator to play this game with a partner or against the computer? What would the tic-tac-toe cards look like?

Teacher's Note:

The Random Number Generator described here does not truly create random numbers but rather simulates them. The random function returns a random integer that is greater than or equal to zero and less than the number specified. On occasion, students may click the Create Number button in the *Hyperstudio* stack and nothing will appear to happen. This would be because the same number was chosen twice in a row. For more information on this topic see page 79 in *HyperLogo-A Scripting Language* published by Roger Wagner.

Tech Time

TIME MARCHES ON

Students will be learning the basic units of time during this activity (seconds, minutes, hours, days, weeks, and years) with the use of calendars and clocks.

Grade Level: three to five
Duration: 60-90 minutes on the computer
Materials: computer, CD-ROM, clock programs, Calendar Survey Question sheet (page 37)
Software: any calendar creating program including *ClarisWorks*, *MSWorks*, *Calendar Creator*, etc.
Internet Link: http://www.c3.lanl.gov/mega-math/welcome.html

Procedure: For Activity #1

Before the Computer:

- Because the first part of this activity involves the use of calendars, students should be familiar with the terms: day, week, month, and year and should review how to read a calendar.

- Students should also know what abbreviations are used for each of the months and the days of the week.

On the Computer:

- There are numerous calendar creation programs available on the market. Programs like *ClarisWorks* and *MSWorks* have everything students need to create beautiful calendars.

- The first calendar that students will create is a monthly calendar. Make it for the current month and year. When using *ClarisWorks 4.0*, the calendar might look something like the monthly calendar above.

- Most calendars will set the week from Sunday though Saturday. A month will either have 28, 30, 31 or 29 days (this happens in the month of February every 4 years).

- On the next page, students will find the Calendar Survey Questions sheet. After both the monthly and yearly calendars have been created, students will be ready to complete the survey sheet.

- The yearly calendar was created using a shareware program called *My Personal Calendar*. Students can create a similar type of calendar using *MSWorks* or *ClarisWorks*. Check the individual instructions for each program.

TIME MARCHES ON (Cont.)

Calendar Survey Questions

Monthly Calendar

1. How many days in a week? _____

2. About how many weeks in a month? _____

3. If Janice were on vacation on the third of the month, and she returned on the 21st, how many days was she on vacation? _____

4. What is the date of the second Sunday on your calendar? _____

5. How many Tuesdays are on your calendar? _____

6. Do any days of the week appear five times on your monthly calendar? What are they? _____

Yearly Calendar

7. How many months make up one year? _____

8. How many weeks make up one year? _____

9. How many days make up one year? _____

10. If Tim left for a sailing trip on April 7 and returned on June 24, how many months and days had he been away on his trip? _____

11. If 24 hours equal one day, how many hours would be needed to equal 17 days? _____

12. If 60 seconds equal 1 minute, and there are 60 minutes in one hour, how many seconds are in one day? _____

Tech Time

TIME MARCHES ON (Cont.)

Assessment:

When assessing this activity, there are two things to look for. First, were both types of calendars created and printed? Second, were the students able to use their calendars to answer the questions on page 37. Questions 11 and 12 will prove to be challenging, and the use of a calculator would be acceptable.

Follow-up:

- Using a reference CD-ROM such as a multimedia encyclopedia, students could do some research on different types of calendars that have been used throughout recorded history. Using a word processor such as *The Writing Center*, they could create mini-reports that could be presented to the entire class.

Procedure: For Activity #2

Before the Computer:

- The second activity works directly with the manipulation of computer clocks. Some computers allow students to change the clock settings to a traditional analog clock and then these changes are updated. If the students' computers do not have this feature, teachers could try either a commercial clock teaching program or try a shareware version. You will need one of these programs to do Activity #2.
- Make sure that students are familiar with the Control Panel and Date & Time settings. They will be using the Control Panel to alter clock settings and then looking at the results on any one of the programs mentioned above.
- Leave both the date and time Control Panel and the clock open. The changes will be displayed on the clock.

On The Computer:

- Open Date & Time found in the Control Panel. Students can usually get to this from the Apple Menu found at the top left corner of the screen. When using a PC, students will find a built-in adjustable clock in both Windows 3.1 or Windows 95.
- In this activity, students will be setting their clocks for different times and recording their answers on the worksheet found on page 39. The time that they will need to create is printed below the clocks. Remember that once a clock is reset it continues to tick so do not take too long to record the answer. Also, when students are finished doing this activity, they should be sure to return the clock to the proper time.

Tech Time

TIME MARCHES ON *(Cont.)*

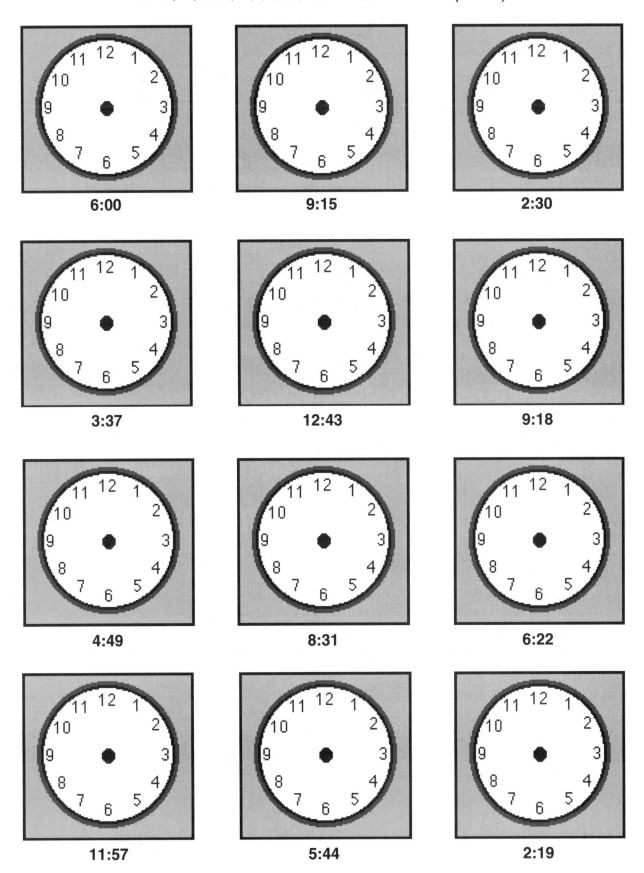

6:00   9:15   2:30

3:37   12:43   9:18

4:49   8:31   6:22

11:57   5:44   2:19

Tech Time

TIME MARCHES ON (Cont.)

Elapsed Time:

- The next two questions involve elapsed time. Use the pairs of clocks at the right to first draw pictures of the two times, then find the answer to the questions.

 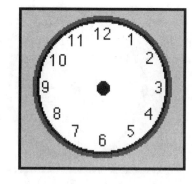

Question-1 Susan went to the store at 3:15 PM to do some shopping. When she arrived back home she noticed that the kitchen clock said 7:47 PM. How much time did Susan spend shopping?

Hours _____ Minutes _____

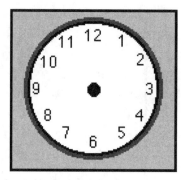 

Question-2 Veronica went to school at 8:15 AM. After school she went to basketball practice. When she finally arrived home, it was already 6:22 PM. How long was Veronica gone?

Hours _____ Minutes _____

Assessment:

The worksheets that were provided for this activity should be useful for assessment. In addition, teachers may want to try some of the follow-up activities. Children at this age seem to have the most difficulty with problems involving elapsed time, so the students may need additional practice.

Follow-up Activities:

- Students can quiz each other by making up their own clock times to be drawn without the aid of the computer.

- Since many students do not have as much trouble reading digital time, a class game could be made out of converting digital time into an analog clock picture.

- There are several activities that could be done to show just what the relative values of standard time units are. What is a minute? What could be done in just one minute? Have students see if they can hold their breath for a minute, or, better yet, "not talk" for a whole minute.

- In a journal, or on a computer program like the *Amazing Writing Machine* or *The Writing Center*, have students list all the things that could be done in a second, minute, hour, day, week, month, year, decade, and century. Compare lists to find those things in common and test some of them.

- Have students design a new kind of clock, either on the computer, or with paper and pencil. They must explain how it works and some of its advantages over the types of clocks currently used in the world.

Dive Into Division

SHOW ME HOW

During this activity students will demonstrate all the steps that are involved in doing division. The finished product will be a multimedia presentation that explains this operation.

Grade Level: three to five

Duration: 60–90 minutes on the computer

Materials: computer, Multimedia Preplanning Sheet (page 43)

Software: *ClarisWorks, MSWorks, Kid Pix Astound, Hyperstudio, HyperCard* or other multimedia software

Internet Link: http://www.astro.virginia.edu/~eww6n/math/math.html
(go to Search and type Division)

Procedure:

Before the Computer:

- In order to be able to do division, students need to understand many other concepts including subtraction, estimation, multiplication, and remainders. The grade level will determine what problems go into this multimedia presentation.

- Students should choose a problem to demonstrate and use the Multimedia Pre-planning Sheet on page 43 to outline each step of their presentations. Narration they will be recording for each frame should be included. Programs like *Hyperstudio* or *Astound* will allow students to also narrate their presentations as part of the production process.

- For younger students working on easier division problems, graphics can be added to each screen. The sample screens below were done in *Hyperstudio*

On the Computer:

- Once students have completed their planning sheets, they can begin creating their presentations. They should show everything that is done in order to solve a division problem, and explain in detail on the soundtrack what each screen means. In addition, there can be an automatic timer so that the stack runs like a slide show.

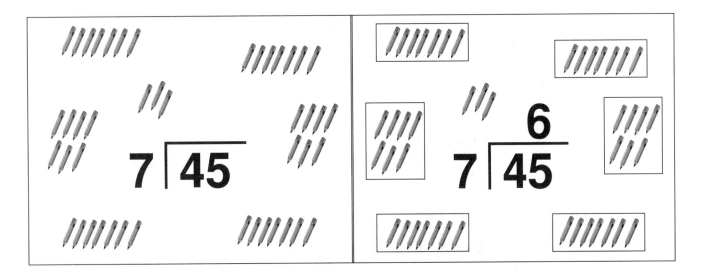

Dive Into Division

SHOW ME HOW (Cont.)

Assessment:

Giving directions is a very complicated skill. Following directions can be just as complicated. One of the ways to test the validity of the program written by students, is to see if another student can explain what he thinks the presentation means. Students may try out their tutorial on younger students to see if they understand the concept. Those projects that have a great deal of detail and correctly state the processes involved in doing division should get a high mark.

Follow-up Activities:

- There are many excellent products on the market that teach division through the use of a personal computer. My personal favorites are those products put out by Edmark. I would highly recommend any of their products.

- Some shareware programs for younger students that teachers might want to try are *D.R. Math*—Mark Starlin and *C C Math 3.0*—Bruce A. Pokras, which are two of my favorites.

Challenge:

- If the students do not already have a computer journal, teachers can find many excellent ones on bulletin boards on the Internet. They can also use programs such as *ClarisWorks* to create their own journals.

- Journals can be used to create division word problems. Challenge students to come up with as many division word problems as they can on the computer. Later these can be used in a problem solving contest. Students must be able to solve their own word problems.

- Some popular shareware journal programs include, *My Journal Notebook 2.1*—MacLondon Creations, *The Journal*—Public Domain, and *JK 2.0*—Randolph Rowan

Dive Into Division

SHOW ME HOW *(Cont.)*

Multimedia Preplanning Sheet - Division Presentation

Slide 1

Spoken Text: _____
_____
_____

Graphics _____

Slide 2

Spoken Text: _____
_____
_____

Graphics _____

Slide 3

Spoken Text: _____
_____
_____

Graphics _____

Slide 4

Spoken Text: _____
_____
_____

Graphics _____

Slide 5

Spoken Text: _____
_____
_____

Graphics _____

Slide 6

Spoken Text: _____
_____
_____

Graphics _____

Slide 7

Spoken Text: _____
_____
_____

Graphics _____

Slide 8

Spoken Text: _____
_____
_____

Graphics _____

Slide 9

Spoken Text: _____
_____
_____

Graphics _____

Finding Averages The Easy Way

NOT YOUR AVERAGE BOARD GAME

During this activity, students will be creating a board game that can be used to learn the process of finding averages. Students will work with rating scales.

Grade Level: three to five
Duration: 90–120 minutes on the computer
Materials: computer, clip art, game pieces, Random Number Generator (Page 28), Not Your Average Board Game Score Sheet (Pg. 46), Assessment Sheet (Pg. 47)
Software: *ClarisWorks*, *MSWorks*, *Kid Pix* or other drawing software
Internet Links: http://www.blueberry.co.uk/thehill/nonshocked/games/index.html
http://www.astro.virginia.edu/~eww6n/math/math.html (go to Search and type Mean)

Procedure:

Before the Computer:

- In the event that some students have never seen a board game before, teachers might want to show them a few samples of popular games.
- During this activity, students will be creating a board game by following **very specific** directions. Part of the assessment will take into consideration how well they followed those directions. Teachers can easily illustrate the importance of step-by-step directions by bringing in directions for assembling many different kinds of items that are sold today unassembled.

On the Computer:

- **Follow each of these steps in order. Try to do them exactly as they are written.**
- On a blank drawing screen, create a square shape that is 5 inches by 5 inches (12.7 cm x 12.7 cm). If the program has an "Auto Grid" function, one that automatically sets the lines at defined intervals, students should use it.
- From each corner draw a straight line 1 inch (2.54 cm) in from the outside edge of the square.
- The shape should now look like figure A.
- At 1 inch (2.54 cm) intervals, students are now going to fill in the shape along the outside edge. The four corners will later become special places.
- Their drawings should now look something like the one pictured here.
- They can add color to the corners by using a Fill Tool. Students will also notice that some graphics have been added in the center of the game board as well as a title.

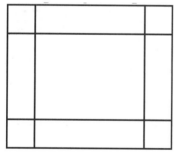

Figure A

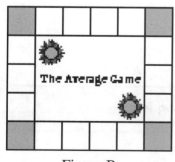

Figure B

Finding Averages The Easy Way

NOT YOUR AVERAGE BOARD GAME (Cont.)

On the Computer: *(Cont.)*

- Now students are ready to add their special places at the corners. Make one corner the Start location. This can be done by placing a large S at this location. Here is what the other three corners will say:

    1. Roll Again - roll the dice or random counter again.
    2. Skip A Turn - skip your turn this time; you cannot score a point.
    3. Earn A Free Point—arn a free point.

- Later students can change these corner titles to something else if they wish. The game board should now look something like this.

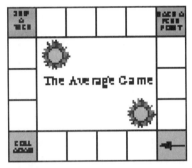

- The last thing that needs to be done is to fill in the numbers that will be used on the game board. If students want to make this board reusable, they should laminate it so that students who use it later can change the numbers if they wish. They will need to use a marker or pen that is erasable.

- Students in 3rd grade may want to stick with only one and two digit numbers, such as 7, 24, 54, 9. Make sure that every blank space has a number. Students in grades 4 and 5 should use numbers with 2, 3, and 4 digits, such as 24, 487, 3,278, 44.

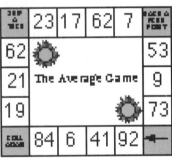

- The board should now look like this, and students will be ready to play once they have printed it out and **READ THE RULES**.

- Students will need either one die (dice) or the random number generator that you created on Page 28. If you are going to use the random number generator, you will need to make one change in the *HyperLogo* script so that you only generate numbers between 0 and 6. Here is the change: **setfieldtext [] "Number Random (7)"**. In order to make this change start up Hyperstudio. Open the stack that contains the random number generator. Select the Button editor symbol and click on the Create Number button. Click on the word Actions and then double click on Use *HyperLogo*. Now make the change.

- If students have not done the random number generator project, they can always use dice. In order to check and score answers, they might want to have either a computer calculator or a desktop one available.

Follow-up Activities:

- Without making any modification, this same activity could be used for addition, number sequence (from greatest to least, from least to greatest), fractions (regular or irregular), and ratio proportion. The numbers will need to be changed to fit the skill on which students wish to work.

- Now that students have some experience in creating a computer math game, see if they can come up with other designs. Then have them write clear directions for the use of their creations.

- Self-assessment, which was used in this activity, is a good way to find out what students think and feel about their work. This kind of strategy can be used in some of the other activities found in this book.

Finding Averages The Easy Way

NOT YOUR AVERAGE BOARD GAME *(Cont.)*

Game Rules:

- Decide who will go first. Using the random number generator or a die, the first player takes a turn and records the number. The first player goes again and records the second number. The first player goes a third time and records the third number. To find the average, the three numbers are added together and then divided by three. The other players perform a check to see if the answer is correct. When using a calculator, students should remember the *rules* regarding the rounding of numbers. If the answer is correct, this is marked on the score sheet. The player doing the averaging cannot use a calculator.
- It is now the second player's turn. Repeat the steps described above. If there are more than two players, each of them will repeat the steps described above.
- If students land on Skip A Turn on any of their three chances to move, they cannot score and must give up their turn to the next person. If they land on Earn A Free Point, they should give themselves a point on their score sheet. If they land on Roll Again, first finish the turn they are on and then go again. If they happen to land on Roll Again while in their second turn, they get to go another time.

Not Your Average Board Game-Score Sheet

Player Name	Scoring Section (Check off one box for each correct answer)
_____ _____	☐☐☐☐☐☐☐☐☐☐ ☐☐☐☐☐☐☐☐☐☐
_____ _____	☐☐☐☐☐☐☐☐☐☐ ☐☐☐☐☐☐☐☐☐☐
_____ _____	☐☐☐☐☐☐☐☐☐☐ ☐☐☐☐☐☐☐☐☐☐
_____ _____	☐☐☐☐☐☐☐☐☐☐ ☐☐☐☐☐☐☐☐☐☐

Finding Averages The Easy Way

NOT YOUR AVERAGE BOARD GAME *(Cont.)*

This is a self-assessment activity. Evaluate your own projects using the matrix below and turn it in to your teacher.

On each of the following items give yourself a score of (0-4)

0 = Did not understand or do it, **1** = Only understood a little, **2** = Understood what you were supposed to do in this activity, **3** = Understood all directions and could do them by myself, **4** = This part was easy for me and I did a great job.

1. How well did you understand what the purpose of this activity was? (0-4) _____

2. In one sentence, tell what you think you were supposed to do in this activity. _____

3. How well did you understand the directions for building the game board? (0-4) _____

4. What score would you give your finished game board if zero was the lowest score and four was the highest score you could receive? (0-4) _____

5. How well do you understand the rules of the game? (0-4) _____

6. Can you explain them in your own words to someone else? (yes or no) _____

© Teacher Created Materials, Inc.   47   #2425 Integrating Technology Into the Math Curriculum

Fractions Can Be Fun

THE FRACTION MACHINE

During this activity students will be investigating the relationships between fractions. They will learn that a fraction represents a part of a "whole" and can be represented as a number.

Grade Level: three to five
Duration: 60–120 minutes on the computer
Materials: computer, clip art, Fraction Screen Chart (Pgs. 49, 51, 53, 55)
Software: *ClarisWorks, MSWorks, Kid Pix, Astound* or other drawing type of program
Internet Links: http://forum.swarthmore.edu/paths/fractions/e.fracdrmathstud.html

Procedure: For Activity #1

Before the Computer:

- Students should have a fair understanding of what a "whole" means. Have them understand clearly that nearly anything can be called one whole, even groups of items like 5 dogs, 3 candy bars or 1 pie.
- In the first part of this activity, students will be making fractions using clip art and their own drawings. Students will create fractions by taking a group of objects and calling them one whole. Then they will separate some of these objects from the whole and say what part of the whole they represent. Teachers should go over the concept of fractions with students before they start their drawings. Students should be clear on the terms numerator and denominator. Fractions are easier to draw if they are grouped by multiples.

On the Computer:

- Students should look at these samples which were done in *Kid Pix*. In the first example, there are a total of 4 apples. Two of these apples have a line drawn around them. This means that the two represents the part and the four represents the whole. This is called two fourths and it is written as 2/4.
- While students are on the computer during this session, they will be creating fractions just like these. Students should label each fraction just as these have been labeled. If they are confused about what to do, they should talk to one of the classroom experts, or ask for the teacher's assistance.
- Students should print their screen results and turn them in with their Fraction Screen Chart.

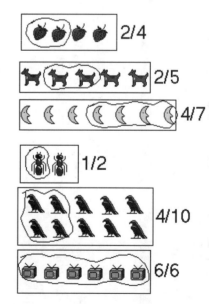

THE FRACTION MACHINE (Cont.)

Fraction Screen Chart- Activity #1

- Create the following fractions on your computer with any drawing program.

2/3	two thirds	5/10	five tenths
1/6	one six	8/8	eight eighths
4/8	four eighths	6/7	six sevenths
3/9	three ninths	4/5	four fifths
2/7	two sevenths	1/3	one third

- Draw each screen using the form below. You should label each screen with the fraction it represents. Turn this chart in with your printout from your computer of the created fractions.

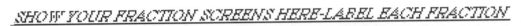

SHOW YOUR FRACTION SCREENS HERE-LABEL EACH FRACTION

Fractions Can Be Fun

FRACTIONS IN A BOX

Procedure: For Activity #2:

Before the Computer:

- For the first activity, students created fractions by taking a group of objects and calling them one whole. Then they separated some of these objects from the whole and said what part of the whole they represented. In this activity, they are going to take one object and divide that object into **equal** pieces.
- If students have ever cut a whole cake or a whole pie into smaller equal pieces, they already know how to make fractions.

On the Computer:

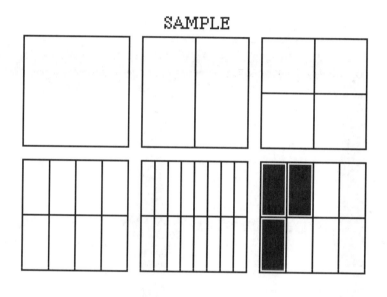

- In the drawing program, students should create a square shape that has a black outline and is hollow in the middle. Divide this shape into halves, fourths, eighths, or the amount necessary. Remember to make each division **equal**. When working with fractions the parts of the whole must be equal. Students should print and turn in work once they have completed this section. If the program that students are using has an Auto Grid function, it should be turned on. This will make it much easier to create and divide the shapes.
- In order to represent fractions, students should shade the part that they wish to be the numerator. In the sample, 3 out of a total of 8 pieces have been shaded. This is called three eighths or 3/8.
- Students should be clear about how to create fractions in this manner.
- Students should create the fractions found on the Fraction Screen Chart on page 51 in the manner just described here. They should not forget to label each fraction, to make each part equal, and to print the screen results when completed.

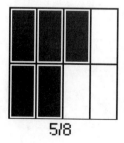  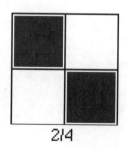  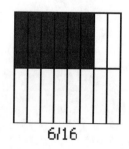

5/8 1/2 2/4 1 6/16

FRACTIONS IN A BOX (Cont.)

Fraction Screen Chart- Activity #2

- Create the following fractions on your computer with any drawing program.

2/3	two thirds	5/6	five sixths
4/7	four sevenths	5/16	five sixteenths
5/10	five tenths	4/12	four twelfths
1/9	one ninth	3/9	three ninths
2/5	two fifths	1/4	one fourth

- Draw each screen using the form below. You should label each screen with the fraction it represents. Turn this chart in with the printout from your computer.

SHOW YOUR FRACTION SCREENS HERE-LABEL EACH FRACTION

Fractions Can Be Fun

FRACTIONS IN A CIRCLE

Procedure: For Activity #3
Before the Computer:

- While it is fairly easy to show fractions using square or rectangular shapes, it is a little more of a challenge to use circles. Once again, students will be creating fractions, using a circle rather than a square or rectangle, as the base image. A few things students should think about as they are doing this section are:

    1. What kinds of fractions are easy to draw using circles as a base image?
    2. What kinds of fractions are more difficult to draw using a circle as a base image?
    3. Can you think of a fraction that is impossible to represent with a circle as the base image?

- Here are a few samples of fractions using a circle as the base image.

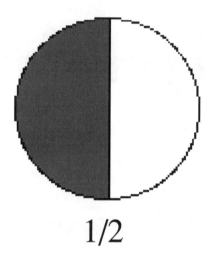
1/2

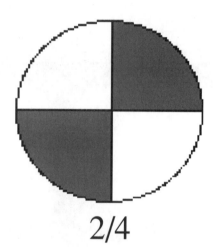
2/4

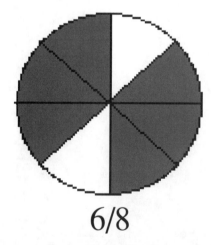
6/8

Fractions Can Be Fun

FRACTIONS IN A CIRCLE (Cont.)

Fraction Screen Chart- Activity #3

- Create the following fractions on your computer with any drawing program using a circle as the base image.

1/4	one fourth	6/6	six sixths
2/8	two eighths	2/5	two fifths
1/3	one third	5/8	five eighths
3/6	three sixths	5/12	five twelfths
3/4	three fourths	2/2	two halves

- Draw each screen using the form below. You should label each screen with the fraction it represents. Turn in this chart with your printout of the created fractions from your computer.

SHOW YOUR FRACTION SCREENS HERE-LABEL EACH FRACTION

© Teacher Created Materials, Inc.　　　53　　　#2425 Integrating Technology Into the Math Curriculum

Fractions Can Be Fun

ALL MIXED UP

Procedure: For Activity #4

Before the Computer:

- So far, students have only worked with fractions that are less than one whole, or regular fractions. It is also possible to have fractions that represent more than one whole, **mixed numbers or irregular fractions**, depending on how they are written. The example below shows a mixed number that represents 2 and 1/4 pies or 9/4. The term "irregular fraction" is used whenever the numerator is larger than the denominator. Teachers should review these concepts before beginning this activity. Teachers can ask students if they can think of a way to turn an irregular fraction into a mixed number.

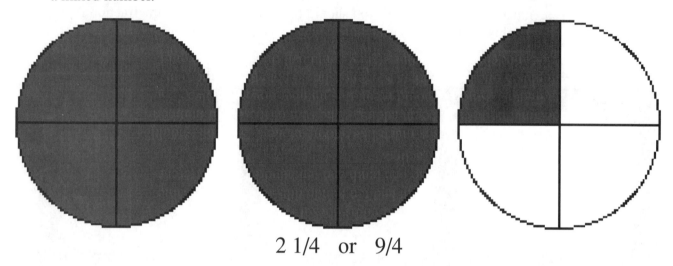

2 1/4 or 9/4

On the Computer:

- During this activity, students will be creating mixed numbers or irregular fractions. Students should start the drawing program and create the following mixed numbers or irregular fractions. They do not need to use a circle as the base image. When they are finished, they should print their computer screen. Students should create the problems on their Fraction Screen Chart - Activity #4 sheet on page 55.
- If students have trouble, they should go to a class expert. Some of these mixed numbers and irregular fractions are easier to display if a rectangular or square shape is used as the base figure.

Follow-up Activities:

- Equivalent fractions are those in which the fractional parts are equal. For example, when students look at the picture for 1/4 and the picture for 2/8, they will see that they represent the same value but have different names. There are many fractions that are equivalent. Students should see how many they can find in a five minute period of time. Record them on a separate piece of paper.

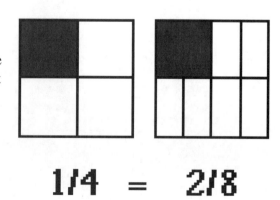

1/4 = 2/8

Fractions Can Be Fun

ALL MIXED UP (Cont.)

Fraction Screen Chart- Activity #4

- Create the following fractions on your computer with any drawing program. Use a circle, square or rectangle base.

2 1/2	two and one half	3 3/7	three and three sevenths
4 2/3	four and two thirds	5/4	five fourths
11/5	eleven fifths	7 2/9	seven and two ninths
5 1/5	five and one fifth	3 1/3	three and one third
2 2/6	two and two sixths	12/4	twelve forths

- Draw each screen using the form below. You should label each screen with the fraction it represents. Turn in this chart with your printout of the created fractions from your computer.

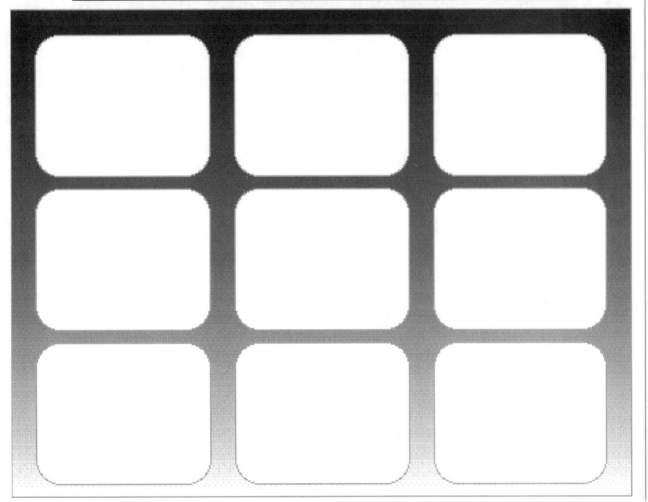

SHOW YOUR FRACTION SCREENS HERE-LABEL EACH FRACTION

Fractions Can Be Fun

ALL MIXED UP *(Cont.)*

Assessment:

The assessment for all activities will consist of the completed form found on pages 49, 51, 53, and 55. Students should also print their screens, if possible. Accuracy is important and, therefore, they should take their time in creating the drawings. One of the hidden objectives is to get students comfortable with drawing, filling, and manipulating objects. While many students can work with pre-drawn stamps such as those found in *Kid Pix*, they often will have much more difficulty creating and manipulating shapes using drawing tools.

Students also need to show that they understand the concept of a fraction being part of a whole. When they create the fractions requested, they show their understanding of the concept. Students can use the form below to test their understanding of how to display all of the types of fractions they worked with in the four activities. This could be used as a graded quiz and done away from the computer or any other kind of aid.

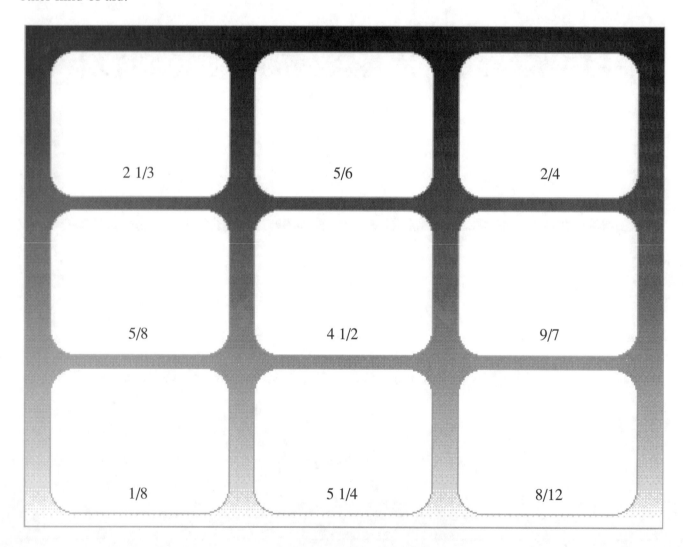

Decimals and Disk Drives Go Together

TRAVELING DECIMALS-TENTHS

During this activity students will explore decimals. They will be creating a destination log that includes a map and how many miles, or parts of a mile, they have traveled.

Grade Level: three to five
Duration: 30–60 minutes
Materials: computer, clip art, Self Examination Assessment Sheet (Pg.60)
Software: any drawing program that has a ruler and grid
Internet Links: http://www.astro.virginia.edu/~eww6n/math/math.html

Procedure: For Activity #1
Before the Computer:

- Students need to be aware of how to use a ruler while working with a computer drawing program. An understanding of how to place a "cross-hair" and how to work with a grid is also important. Many commercial programs allow a great deal of flexibility in the control of rulers, measurement units, and grid placement.

- A basic understanding of tenths, hundredths, and thousandths, and how these relate to fractions is also needed.

- A basic understanding of an odometer and how it works is also needed. Every time the car travels 1/10 or 0.1 of a mile, the number farthest to the right changes. On a car, an odometer might look like this. This car has traveled 45.6 miles not 456 miles. The six stands for six-tenths, or 0.6, miles.

On the Computer:

- During this activity, students will be creating computer-generated trip logs. They can take their trip anywhere in the world, but the maximum distance that they can travel in the first part of this lesson is *only one mile*. They can only travel in straight lines and turn at right angles. They must be very precise in how they draw their trip logs, making sure to follow all directions.

- Students will be completing their trips from a starting point, then back to start, moving at 0.1 mile intervals.

- Students should set the ruler scale to inches in the drawing program. To set the scale, make sure that the ruler function is on and visible. Students may start at any location that they wish on the page, but can only move north, south, east, and west in a straight line.

- **Only right angle turns are allowed.**

- Students should add a few graphics and clearly mark, in decimal form, all distances.

© Teacher Created Materials, Inc.   57   #2425 Integrating Technology Into the Math Curriculum

Decimals and Disk Drives Go Together

TRAVELING DECIMALS-TENTHS *(Cont.)*

On the Computer: *(Cont.)*
- This sample screen was done using *ClarisWorks*. Notice that the rulers are clearly visible and that exactly one mile has been traveled.

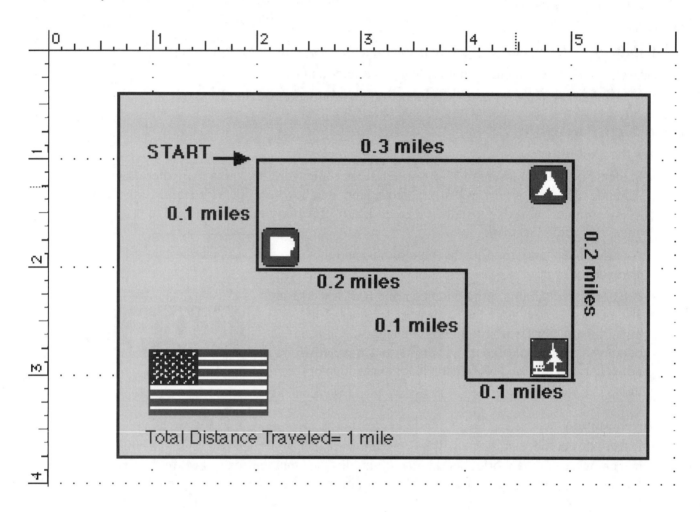

- There are numerous ways that students could draw this kind of map, but they must return to the starting point. Students should begin creating as many trip logs, or maps, as they can. How many maps are possible? More than ten?
- If a printer is available, be sure they print each map they design. They might want to add graphics and a title to their maps.

Assessment:

This activity is working on a number of math concepts simultaneously. The student should have followed each step of the instructions. The total number of tenths should be equal to one mile. The scale of the drawing should be accurate to the nearest inch.

Follow-up Activitiy:
- This activity lends itself very well to concepts in geometry, such as perimeter, area, and measurement, with a little bit of modification.

TRAVELING DECIMALS-HUNDREDTHS

Decimals and Disk Drives Go Together

Procedure: For Activity #2

Before the Computer:

- This second activity is similar to the first, but now students are going to work with hundredths. It takes 10 one-hundredths to equal one-tenth. If students do not understand this idea, they should use the pictures below that illustrate the concept. How many hundredths does it take to equal one-tenth? They can see, both one tenth and ten hundredths represent the same part of the shape shown below. So one tenth and ten hundreds are different names for the same decimal or fraction.

- Go over adding decimals. This is much like adding money. Students should line up the decimal points and begin adding. The sum of these four numbers is fifteen hundredths. Students could break this number down into two parts- 1 tenth plus 5 one hundredths. Students should note that the zero in front of 0.1 does not change the value. They could call this .1 and it would be correct.

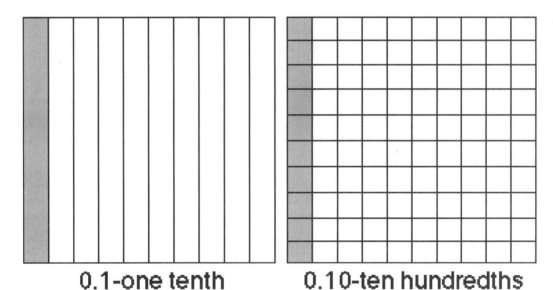

On the Computer:

- In the last activity, students worked with tenths to create a trip log. They are now going to work with hundredths but in a slightly different way. First, each 1/2 inch will equal one one hundredth (0.01). This is a change from the previous activity. Second, they must put together enough hundredths to *equal five tenths* (0.5). As in the first activity, students must move in a straight line and can only make right angle turns.

- Students should label the intervals and make sure that the ruler is visible while they are working. This is not a guessing activity but rather, it requires a high degree of accuracy.

- Have students study the sample on page 60 very carefully before they start. They should notice that each interval is carefully marked and that the total number of intervals equals 0.50 or fifty-hundredths This also represents five tenths 0.5. They may not exceed 0.50 nor can the intervals be less than this amount.

Decimals and Disk Drives Go Together

TRAVELING DECIMALS— HUNDREDTHS (Cont.)

SAMPLE OF ACTIVITY 2

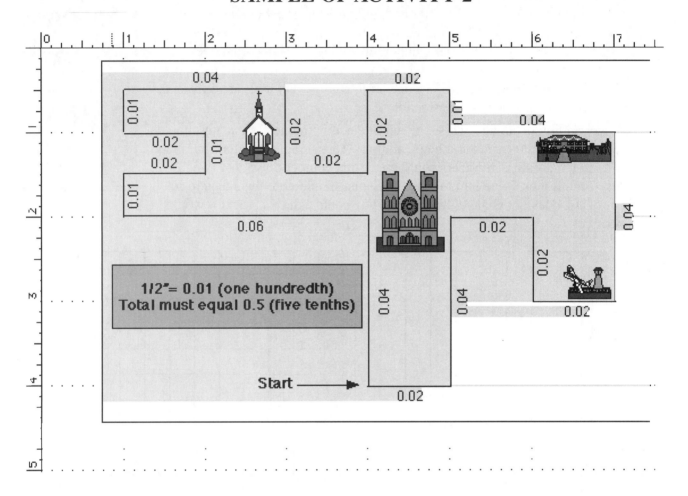

On the Computer: *(Cont.)*
- There is a great deal of detail in this drawing. The students' drawings should reflect the same kind of attention to detail. Many shapes are possible that will add up to 0.5.
- When students are finished, produce a printout of the travel log. They will not be able to print the rulers unless you use a screen capture utility such as *PICTify* by Scott Johnson. Using such a utility will make it easier to evaluate their work.

Assessment:

After working with tenths and hundredths during the past two activities, students should now be able to demonstrate what they understand about decimals. Page 61 provides a written assessment where students will answer questions and do some illustrations by hand to show their understanding. This assessment is to be done individually and without the aid of a computer or any work that was previously done on the computer regarding decimals and fractions. The teacher may want to place a time limit on the completion of this self test. Students will need page 61, a pencil, and crayons or colored pencils.

Decimals and Disk Drives Go Together

SELF-EXAMINATION

Part One:

Answer each of the following questions to the best of your ability.

1. Which is larger, a tenth or a hundredth? _____

2. How many tenths does it take to equal one? _____

3. How many hundredths does it take to equal one? _____

4. What is the device on an automobile that measures in tenths of a mile called? _____

5. (Tough question) Why do you think you were not allowed to use diagonal lines in either of these activities? _____

Part Two:

This section of the assessment involves matching a decimal to a fraction. Draw a line from the decimal to the fraction that matches it on the other side. Careful, **one** fraction will not be used.

1. 0.57		a.	98/100
2. 0.2		b.	67/100
3. 0.78		c.	3/10
4. 0.5		d.	5/10
5. 0.25		e.	57/100
6. 0.3		f.	4/10
7. 0.18		g.	2/10
8. 0.98		h.	18/100
9. 0.4		i.	25/100
		j.	78/100

Part Three:

In this section you will need to draw an illustration for each decimal.

(0.4)   (0.23)   (0.62)

Probability-Nothing More Than Chance

TAKE A CHANCE

During this activity, students will be working with probability. The computer will be used to generate a statistical picture of the results from probability tests that they perform.

Grade Level: three to five
Duration: 30-60 minutes
Materials: computer, tiles, bag, coins
Software: *ClarisWorks*, *MSWorks*, or other spread sheet program
Internet Links: http://www.irony.com/igroll.html

Procedure: For Activity #1
Before the Computer:
- What is probability? This term could be defined as the chance that an event will occur. A coin has two sides, usually called heads and tails. If a student tosses a coin into the air, the chance of getting heads or tails is 1 out of 2. This is because there are two sides to a coin and only one side is heads or tails. This seems simple enough, and yet many people will answer the following question incorrectly.

A Question:
- John picks up a coin and begins to flip it in the air twenty times in a row. Each time he records the result. For the first 20 flips of the coin, **all** tosses resulted in heads. Will there be a greater chance of getting heads or tails on the 21st flip? Look for the answer to this question on page 63.

On the Computer:
- Together with a partner, students are going to flip a coin a total of 50 times. Each time the coin is flipped, the result will be recorded using a simple spreadsheet. See the example. (Predict your results)
- There are several important things that they should notice about this spreadsheet. First, each trial can only have one answer, either heads or tails. Second, the total number of trials must equal the total results for heads and tales (6+4=10). Third, they are going to use a formula or a function to help them calculate the results. Students should check the software manual index for formulas or functions to see how these work properly.
- The word SUM tells location B12 to add everything between B2 and B11. The spreadsheet should have enough room for 50 trials.

	A	B	C
		Heads	Tails
1			
2	Trial-1	1	
3	Trial-2		1
4	Trial-3	1	
5	Trial-4	1	
6	Trial-5		1
7	Trial-6	1	
8	Trial-7		1
9	Trial-8	1	
10	Trial-9	1	
11	Trial-10		1
12	Total	6	4
13			

B12 =SUM(B2..B11)

Probability-Nothing More Than Chance

TAKE A CHANCE (Cont.)

Assessment:

How close were the actual results to what the students had predicted would happen? If they did this a thousand times, what would they expect the results to be? Is there any way that students can think of to make more heads than tails appear?

Answer to Question:

- The answer to the question on the previous page, about John flipping a coin 20 times and getting heads 20 times in a row, might be surprising. On the 21st flip, the odds would still be 50/50 or 1 out of 2 chances that he would get a head or a tail. The previous results do not make it any more likely it will be heads or tails on the next flip.

Procedure: For Activity #2

Before the Computer:

- The next activity involves a few more variables than the previous one. Some of the concepts discussed here could be applied to games of chance such as cards and dice.

- For this activity, students will need colored squares or manipulatives such as those found in the math series. Four or more colors would be a good start. Once again, they will be creating a spreadsheet that uses the same function or formula they used in the last activity, so they should understand how to do this.

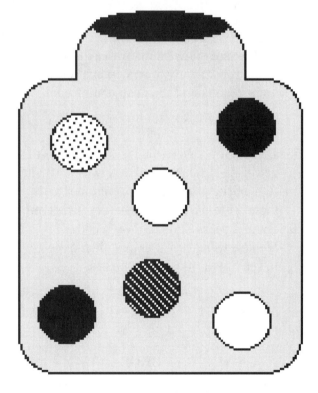

- This is the probability example with which the students will be working. In this drawing, they can see two black marbles, two white marbles, one spotted marble, and one striped marble. Students will be working with colors such as blue, yellow, and red.

- If they can not see into the jar, and stick in a hand to grab any marble, what will be their chance of getting a black marble? The answer is 2 out of 6 or 2/6. Since there are six marbles altogether and two of them are black, the probability of getting a black one is 2 out of 6. The same would be true for the white ones.

- There is only one each of the spotted and striped marbles. The probability of getting one of them would only be 1 out of 6 or 1/6. Notice that this is a fraction.

- Students should have manipulatives in at least four different colors. They should put four red, three green, two blue, and one yellow manipulatives into a paper bag. They may substitute different colors, but in the same quantities given. This will give students a total of ten manipulatives in the paper bag.

- They should draw them out one at a time, and then replace them afterwards so that there will always be ten in the bag before they make the next draw.

Probability-Nothing More Than Chance

MANIPULATIVE MADNESS

Before the Computer: *(Cont.)*

- Students will need to answer these questions before starting:

    a. What is the probability of drawing a yellow one? _____

    b. What is the probability of drawing a blue one? _____

    c. What is the probability of drawing a green one? _____

    d. What is the probability of drawing a red one? _____

    e. Which one do you think you will draw most often? _____

    f. Which one do you think you will draw least often? _____

- One student will draw first while the other records the results, and then they will change turns. They should not peek into the bag while doing this activity.
- After they have done 100 trials, they can do an analysis of the results to see if the probability of choosing a particular color held up. The more trials that are done, the closer the answers will match what probability says should happen. One hundred trials should give them a fair representation of the odds.
- Below are some charts for each of the four colors. These charts show what should have happened after 100 trials with an explanation of how these figures were determnined. Students will record their results and then use math to put it into its simplest form.

RED

100 attempts were made. The chance of getting a RED manipulative was 4/10 since four of the manipulatives were red and there were ten total. For every 10 attempts, about 4 picks should have been red. If there were 100 attempts, then expect to see about 40 picks to be red.

40-projected _____actual

GREEN

100 attempts were made. The chance of getting a GREEN manipulative was 3/10 since three of the manipulatives were green and there were ten total. For every 10 attempts, about 3 picks should have been green. If there were 100 attempts, then expect to see about 30 picks to be green.

30-projected _____actual

BLUE

100 attempts were made. The chance of getting a BLUE manipulative was 2/10 since two of the manipulatives were blue and there were ten total. For every 10 attempts, about 2 picks should have been blue. If there were 100 attempts, then expect to see about 20 picks to be blue.

20-projected actual_____

YELLOW

100 attempts were made. The chance of getting a YELLOW manipulative was 1/10 since one of the manipulatives was yellow and there were ten total. For every 10 attempts, about 1 pick should have been yellow. If there were 100 attempts, then expect to see about 10 picks to be yellow.

10-projected _____actual

Probability–Nothing More Than Chance

MANIPULATIVE MADNESS *(Cont.)*

On the Computer:
- They will now need to construct a spreadsheet like the one in the last activity. Look at the example below. There will be 100 trials so make sure the spreadsheet will allow for this. Notice that this spreadsheet also uses the same formula that was used in the other activity. This time there are four totals for B,C,D, and E. Once again the sum of these totals equals the number of trials.

untitled (SS)

C12   =SUM(C2..C11)

	A	B (Blue)	C (Green)	D (Red)	E (Blue)
1		Blue	Green	Red	Blue
2	Trial-1			1	
3	Trial-2		1		
4	Trial-3		1		
5	Trial-4	1			
6	Trail-5			1	
7	Trial-6		1		
8	Trial-7			1	
9	Trial-8	1			
10	Trial-9			1	
11	Trial-10				1
12	Total	2	3	4	1
13					

Assessment:

How did the students' actual results look when compared to the projected results? An easy way to compare the results is to make a ratio. Suppose they recorded a number of 13 for the yellow manipulative. They now know that yellow should have been picked about 10 times out of 100 trials. The result is a little high. If they turn these both into fractions, it is a little easier to see. 10/100 or 1/10 represents the projected result, and 13/100 or 1.3/10 represents the actual result. Can students think of any reasons why a particular color was picked more frequently than it should have been? What would happen if they changed the ratio of colors in the bag? Why would the results be different?

Probability-Nothing More Than Chance

MANIPULATIVE MADNESS *(Cont.)*

Assessment Activity:

- With a partner, have students perform the following test after they have made their predictions.

- In a bag, put manipulatives of three different colors in the following proportions:
    Color #1- 9 manipulatives
    Color #2- 3 manipulatives
    Color #3- 2 manipulatives

Predictions:

- What is the chance of getting color number one (express as a fraction)?

- What is the chance of getting color number two (express as a fraction)?

- What is the chance of getting color number three (express as a fraction)?

- Decide on how many tests to perform.

- Set up the spreadsheet just like in the last two activities.

- Fill in the chart below with projected results and actual results.

- How did the results compare to the predictions? Were any of the results very different from what was predicted?

COLOR #1 _____   COLOR #2 _____   COLOR #3 _____

Number of Tests _____   Number of Tests (same as #1.) _____   Number of Tests (same as #1) _____

Projected Result _____   Projected Result _____   Projected Result _____

Actual Result _____   Actual Result _____   Actual Result _____

Measuring Matters

WORKING WITH RULERS

During the following activities, students will be working with measurement, perimeter, and area. They will be creating their own metric rulers and measuring a number of items commonly found in the classroom.

Grade Level: three to five
Duration: 60–90 minutes
Materials: computer, rulers, Measuring My Stuff Sheet (Pg. 69), Finding the Perimeter Sheet (Pg. 70), Finding the Area Sheet (Pg. 72)
Software: _ClarisWorks_, _MSWorks_, or other drawing program
Internet Links: http://www.svusd.k12.ca.us/curriculum//Eighth/Tech8/math.html

Procedure:

Before the Computer:

- If students have a printer and a drawing program, they will be able to create their own metric ruler for use in the following activities. If they do not have a printer, they can still do the activities using a metric ruler.

- Students should have an understanding of the difference between metric units of measurement and standard measurement which is still used in the United States. Stress that the United States is one of only a few countries still holding on to this older way of measuring, and that the metric system, as a whole, is much easier to work with.

- Since the very first activity is to create a metric ruler, it would be a good idea to have a printer connected to your computer that will print exactly what is seen on the computer screen. Many older printers are not capable of doing this. Generally, any ink jet or laser jet printer is capable of printing screen captures.

On the Computer:

- Students should start the drawing application and turn on the ruler function. If they are using _MSWorks_, they can find the ruler function in the View pull down menu.

- They will need to change the measurement units on the ruler. The drawing page will be in inches since this is the unit most commonly used. If they are working with _ClarisWorks_, they would find Rulers under the Format pull down menu.

- **Set the units to centimeters.** If students get stuck at this step and do not know where to make these changes, ask one of the classroom experts for help.

- They should now see rulers in centimeters on both the top and left or right edge, depending on which program is being used.

- Most drawing programs, have a View button. This button, which can be found in a menu, will allow students to shrink or enlarge the screen. Shrink, or scale, the picture to 50%. This will allow them to see the entire page and that will make drawing the ruler much easier.

- They are going to be drawing a ruler that is 4 cm, or centimeters, wide and 20 cm long. An example of how this is done is on page 68.

Measuring Matters

WORKING WITH RULERS (Cont.)

On the Computer: *(Cont.)*

- This screen shot shows the three steps needed in order to make a metric ruler. First, students should draw a rectangle that is 4 cm wide and 20 cm long.

- Next add divisions that are 2 cm apart. If the program they are using has an autogrid function, use it to draw the lines.

- Finally, draw lines at the 1 cm interval. They will notice that on the left hand ruler, the first number found is 4. They will also see that on the ruler they are creating, there are 4 cm when they reach this point.

- After they have completed marking off all the 1 cm intervals on the ruler, they can label it. This can be done either before printing or after printing.

- Student rulers should measure exactly 20 cm. They may want to check it against another metric ruler in the classroom.

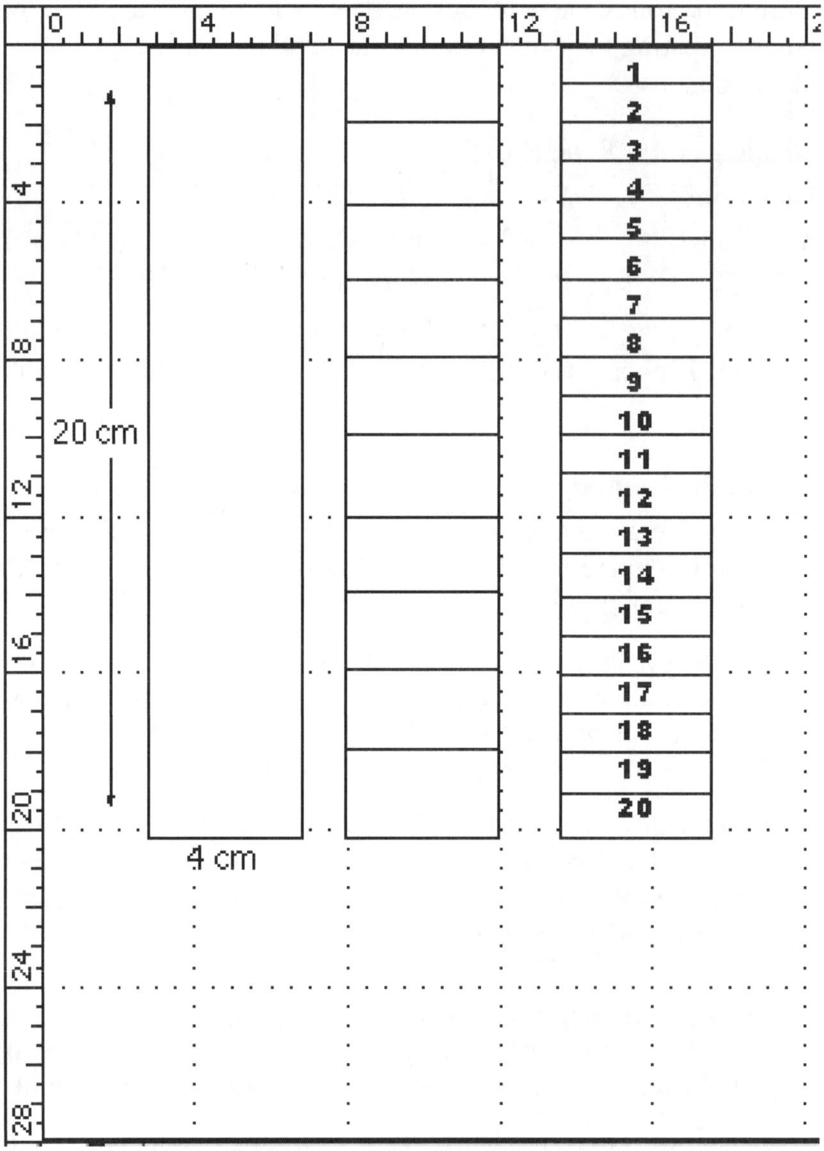

Finding the Measurement:

- Students are now ready to measure a variety of items in the classroom. The following is a list of items that they will be measuring. Before starting, make some predictions about how many centimeters these items are. Use page 69 for the predictions. Students may want to devise a plan that they can use to measure items that are more than 20 cm long.

    a. their math books   b. their pencils   c. the top of their desks

    d. their little fingers   e. the chalkboard   f. their shoes

    g. any ball   h. a crayon   i. a piece of binder paper

#2425 Integrating Technology Into the Math Curriculum   © Teacher Created Materials, Inc.

Measuring Matters

MEASURING MY STUFF

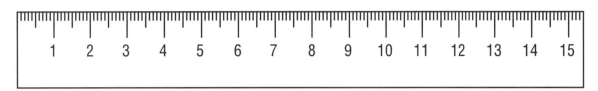

- Use the ruler that you created to measure these objects. Fill in the chart below as you go along. When you are done measuring all of the objects, fill in the assessment questions.

  a. your math book b. your pencil c. the top of your desk
  d. your little finger e. the chalkboard f. your shoe
  g. any ball h. a crayon i. a piece of binder paper

Item	Predicted Size	Actual Size
a. your math book		
b. your pencil		
c. the top of your desk		
d. your little finger		
e. the chalkboard		
f. your shoe		
g. any ball		
h. a crayon		
I. a piece of binder paper		

Assessment:

This is a self assessment activity. You should answer the following questions about this activity.

1. What part of the activity did you find most difficult? _____
   _____

2. How many of the items were larger than 20 cm? What did you have to do in order to measure these items?
   _____
   _____

3. Do you think that everyone doing this activity will come up with the same kind of results that you did? Why or why not?
   _____
   _____

© Teacher Created Materials, Inc.

Measuring Matters

FINDING THE PERIMETER

Procedure: For Activity #2:

Before the Computer:

- Since this activity uses some parts of the first activity found on pages 67–68, it is important that students either do Activity 1 first, or understand what was done.

- In Activity 2, they will be working with the concept of perimeter. They may have already studied this in math class, but teachers should give a quick review of what the term perimeter means. For the purpose of this activity, students will only be working with shapes that have straight lines for sides. Shapes, like the square and rectangle, allow for shortcuts to find the perimeter because some or all of the sides are equal. Students should notice that only one measurement was needed for the square, since all sides are equal. In each sample pictured here, the sides were measured and then totaled.

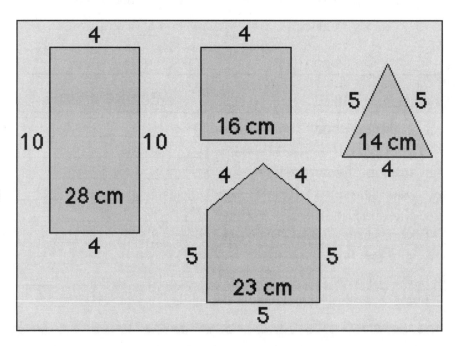

- The examples show students how to calculate the perimeter of a number of different shapes. In the following activities, they will not only be measuring the perimeter of items, but they will also be creating shapes on the computer with a specific perimeter.

- The unit of measurement that was used here is the centimeter, the same unit that students will be using.

On the Computer:

- Since some of the objects or spaces students will be measuring will be greater than 20 cm, they are going to enlarge their rulers. If students have forgotten how to make a 20 cm ruler, they should go back to page 68.

- How many centimeters equal one meter? How many 20 cm of rulers will be needed to make a ruler that is 100 cm long? Students can print five 20 cm rulers and then tape them end-to-end. They can label the intervals in any way that works for them, either before printing the 20 cm rulers or after. Teachers should point out that one meter is just a few inches longer than a yardstick or three feet.

Finding the Perimeter:

- When students have completed the steps above, they are ready to find the perimeter of some commonly found objects or places. They should fill out the form on page 71 with their predictions and actual results. If students are not clear about what they are supposed to do, have them ask one of the classroom experts for help.

Measuring Matters

FINDING THE PERIMETER (Cont.)

- Use the ruler that you created to measure these objects. Fill in the chart below as you go along. When you are finished measuring all of the objects, fill in the assessment questions.

    a. your classroom
    b. your desk
    c. the chalkboard
    d. the basketball court
    e. your chair
    f. your reading book
    g. your teacher's desk
    h. your foot
    i. a piece of paper
    j. an object of your choice

Item	Predicted Size	Actual Size
a. your classroom		
b. your desk		
c. the chalkboard		
d. the basketball court		
e. your chair		
f. your reading book		
g. your teacher's chair		
h. your foot		
I. a piece of paper		
j. your choice		

Assessment:

This is a self-assessment activity. You should answer the following questions about this activity.

1. What part of the activity did you find most difficult? _____
   _____

2. How many of the items were larger than one meter? What did you have to do in order to measure these items?
   _____
   _____

3. Do you think that everyone doing this activity will come up with the same kind of results that you did? Why or why not?
   _____
   _____

Measuring Matters

FINDING THE AREA

Procedure: For Activity #3:

Before the Computer:

- In the previous activity, students learned that they could find the perimeter of an object by adding up the distance around that object. In this final activity, they will find the area of some of the items from the previous list. In order to do this, they will once again need the 100 cm or 1 meter ruler.

- With students, go over the concept of finding the area of a space. In the examples below, some of the shapes require that they multiply more than once and then add the results together. This is how the area of a home would be measured.

- These three examples may help students with the concept of how the area of a shape is found. The area is the space within a shape.

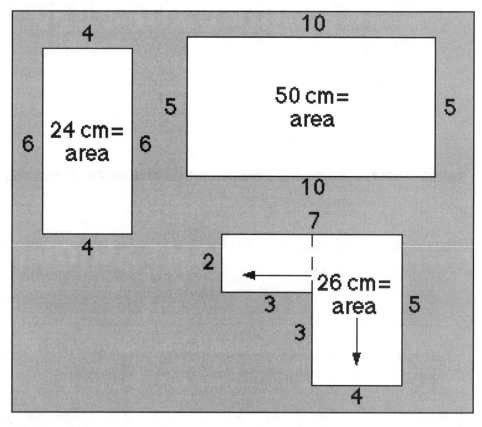

- In order to find the area of a square or a rectangle, students should multiply one side by the adjacent side. This will tell them how many square centimeters the shape is. They would label this answer 50 sq. centimeters.

- The third shape is a little different. There are two different rectangular parts to this shape, and one way students can find its area is to find the area of each part and then add the two together. This is what was done to find the area of this shape. First, the smaller left part was found to be 3x2= 6 sq. cm. Then the larger right part was found to be 5x4= 20 sq.cm. When added together, 20 sq. cm and 6 sq. cm, total 26 sq. cm.

- Shapes such as triangles and those with irregular sides, such as a pentagon, are more difficult to calculate the area.

Measuring Matters

FINDING THE AREA (Cont.)

Finding the Area:
- Find the area of some of the shapes with which you have previously worked. If you did not do the previous two activities, then you will need to take measurements on these items first.
- Find the area of the following shapes or spaces:

    a. your desk
    b. the blackboard
    c. a piece of paper
    d. your classroom
    e. one of the books in your desk

ITEM	PREDICTED	ACTUAL
a. your desk		
b. the blackboard		
c. a piece of paper		
d. your classroom		
e. a book		

Assessment:

This is a self-assessment activity. You should answer the following questions about this activity.

1. What part of the activity did you find most difficult? _____
_____
_____

2. Do you think that everyone doing this activity will come up with the same kind of results that you did? Why or why not?
_____
_____
_____

Follow up Activities:
- Is there a way that you can find the area of a triangle using measurement? Here is a hint. Study this little drawing for a minute.
- Some shapes have the exact same number for the perimeter and the area.
- An example would be a square whose sides are 4 cm. This would give it a perimeter of 16 cm and an area of 16 sq. cm. Can you think of any other examples where this is true?

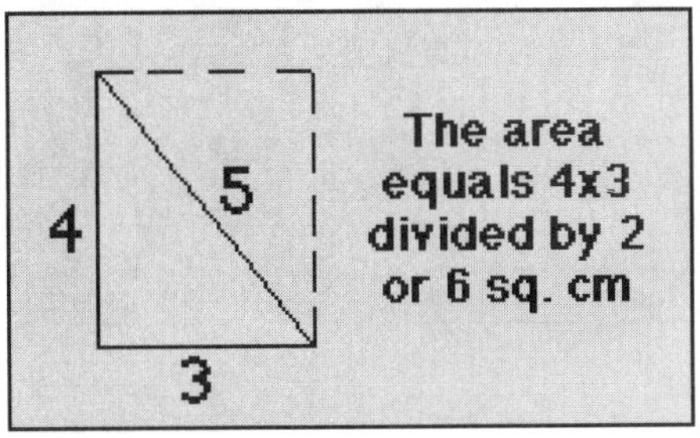

The area equals 4x3 divided by 2 or 6 sq. cm

Going Geometric

CROSSWORD GEOMETRY

Students will be working with elementary geometric shapes to create a crossword puzzle about these shapes and then put together a multimedia presentation on geometry.

Grade Level: three to five
Duration: 60–120 minutes
Materials: computer, printer, Hyper Geometry Sample Stack (Pg. 76), and Hyper Geometry Assessment Form (Pg. 79)
Software: a crossword puzzle creation program, *Hyperstudio 3.0,* or other authoring program
Internet Links: http://freeabel.geom.umn.edu/

Procedure:

Before the Computer:

- Students will need to be familiar with crossword puzzles and how they work. Most children this age have had exposure to crossword puzzles, but may have never tried to make one. There are many different kinds of commercial crossword puzzle makers available or teachers may find a shareware product..
- Students should be familiar with the following geometric terms: square, rectangle, pentagon, hexagon, octagon, circle, triangle, and vertices.

On the Computer:

- Using the words square, rectangle, pentagon, hexagon, octagon, circle, triangle, and vertices, students will create a crossword puzzle. The examples that follow were done on a shareware program called *Cruciverbalist PPC* by the West Pole Software Company. Students might also want to try *CMaster 0.4* by Patrick Stevenson. Some crossword programs will automatically generate the puzzle from the word list that is entered. Some allow students to manually place words wherever they choose.

Across Clues	Down Clues
14 3 sides and 3 vertices	1 4 sides and 4 vertices
15	2
16	3
17 6 sides and 6 vertices	4
18	5
19	6
20	7
21	8 4 equal sides and 4 vertices
22	9
23 5 sides and 5 vertices	10
24	11 8 sides and 8 vertices
	12
	13

Going Geometric

CROSSWORD GEOMETRY *(Cont.)*

Assessment:

Are other students able to solve the crossword puzzle that the students created? Remember that in a crossword puzzle all connections, both horizontally and vertically, must be words. Is this true of the students' crossword puzzles?

Follow-up Activities:

- Create a new crossword puzzle using these new related terms. If students do not know what the terms mean, they should look them up or ask a classroom expert.

Challenge Terms:

a. ray
b. angle
c. line
d. line-segment
e. symmetry
f. congruence
g. perimeter
h. area
i. pyramid
j. cube
k. cone
l. sphere
m. prism

Procedure: For Activity #2:

Before the Computer:

- In this activity, students will be using *Hyperstudio* or any other authoring program to create a multimedia presentation that demonstrates their understanding of basic geometric terms.

- Students will be adding buttons and graphics, as well as drawings, to each card in their stack. Have students look at the example screens found on page 76. These screens show how students' screens should look. They will also be adding some testing functions in order to allow others using their stacks to know if they have answered the geometric questions correctly.

- When they have completed their stacks, they can publish the programs either school-wide or on the Internet, to challenge others.

Going Geometric

HYPER GEOMETRY

Sample Stack

- These screen shots should present a clear idea of the nature of this project. The next page will describe in more detail how you will generate these cards using the tools found within *Hyperstudio* program.

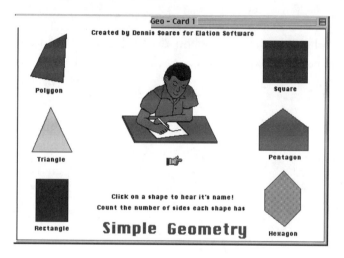
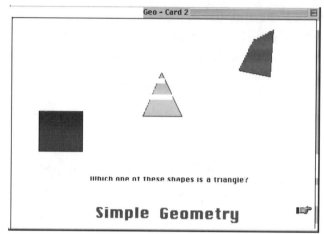
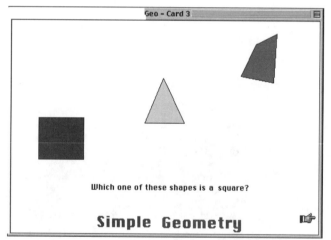
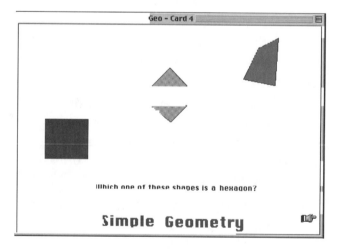
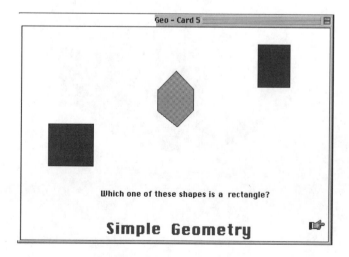
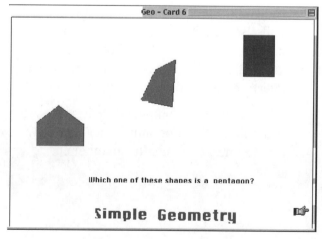

Going Geometric

HYPER GEOMETRY *(Cont.)*

On the Computer:

- Students will be creating a stack that consists of 10 cards using *Hyperstudio* or any other authoring program. An example of the layout of the cards in the stack is shown on page 76. The following instructions are for using *Hyperstudio,* but they are very easily adapted to any program.

- After students have studied the sample screens, they will find instructions on how to create these screens using the least amount of effort.

- Screen #1: This is a simple introduction screen. Students can use the drawing tools to create the shapes. A graphic image, the boy at the desk, was placed on the card using the Add A Graphic Object from the Object menu. Each of the geometric shapes was turned into a button by choosing Add A Button from the Objects Menu. Once students are here, use this tool to select the entire shape as a button.

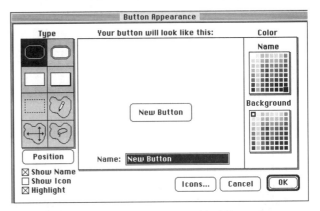

- When they get to the Things To Do—Places To Go Window, they will be doing several things. First, for each shape, students will record the sound for that shape. In order to do this, click on the Play A Sound option and record the proper sound for that shape. Do this for all of the shapes on the opening screen. Remember to add a title and the name of the author of the stack.

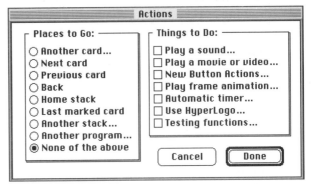

- There is one more button to add to the opening screen, and it is found on the bottom right hand corner. This button is an icon and takes the user to the next screen. Add this button, then under Actions, tell it to go to the next card.

- Students are now ready to move on to the rest of the program. Since they have already created all the geometric shapes they need for the first six screens, they can use some shortcuts to do these screens.

Using Copy And Paste Functions:

- The cut, copy and paste functions found under the Edit menu, are much like that of the word processing functions.

- From the Edit menu choose New Card. Students can now add a different background color or use a gradient to fill.

- Students should go to Move and move to the opening screen. Change the tool to either the Button editor or the arrow. Click once on any one of the geometric shapes. This selects the button. Students can now move to the Edit menu and choose Copy, move to card #2 and select Paste from the Edit Menu. Students should repeat this sequence of steps until they have 3 geometric shapes, that are now buttons on card #2.

© Teacher Created Materials, Inc. 77 #2425 *Integrating Technology Into the Math Curriculum*

Going Geometric

HYPER GEOMETRY (Cont.)

On the Computer: *(Cont.)*

- At the bottom of the card, pose a question such as, "Which one is a triangle?" Make sure that the answer is on the card!

- The last operation students should do with this card is to change the sounds on the incorrect answers. Again, change to the [B] or arrow in the Tool menu and select, or double click, an incorrect shape. Choose Actions and then double click on the Play a Sound option. Change the sound to an incorrect response sound or record one of your choice.

- When students have Card 2 completed, they can go to the Edit menu and choose Copy Card, then choose New Card, then Paste Card. Perform this function four times.

- They will now need to reorganize the shapes, copy some additional ones from the opening screen, and change the questions. Students should create their own cards for cards 3 through 6.

- Card 7 represents a change, although the next page button at the bottom of the screen remains the same. They will need to draw a line, line segment, and ray on this card. Using the same technique they used earlier, make each drawing into a button. After turning each of these three drawings into a button, they will need to record the proper sound that goes with each one. Cards 8, 9, and 10 are identical to card 7 except that the question changes on each one. One simple way to do this is to copy Card 7 and paste it three times.

- On Cards 8, 9, and 10, students will need to make sure that the sounds for the incorrect answers are also changed.

Adding Automation:

- Students can easily automate this stack so that every time the user clicks on a correct answer, it not only plays the correct sound but also moves to the next card. Since a button can do multiple things when it is clicked, they can also tell all of their correct answers to Go To Next Card when they are chosen.

Assessment and Follow-up:

- The assessment form for this activity is found on page 79. Students will not be assessing their own work. Three other students will be evaluating the work and rating it on a scale of 1-10, based on a number of criteria. *Each student* will need to have three other students evalute his or her work.

- When evaluating other projects, students should try to be fair. They should honestly look at each item on the checklist and give it a score that represents the work that was done.

- Teachers should be sure that students understand what they are looking for in the evaluation process. If they are unclear about how to evalute, they should ask one of the classroom experts.

HYPER GEOMETRY *(Cont.)*

Hyper Geometry Assessment Form

QUESTIONS: (Score 1-10)   **SCORE**

The stack has a minimum of 10 cards. _____

The "Next" button works on all cards. _____

The sounds are appropriate on all cards. _____

Rate the appearance of the stack. _____

The stack is easy to use. _____

The stack teaches basic geometry facts. _____

Rate the opening screen. _____

Total Score ⟶

Hyper Geometry Assessment Form

QUESTIONS: (Score 1-10)   **SCORE**

The stack has a minimum of 10 cards. _____

The "Next" button works on all cards. _____

The sounds are appropriate on all cards. _____

Rate the appearance of the stack. _____

The stack is easy to use. _____

The stack teaches basic geometry facts. _____

Rate the opening screen. _____

Total Score ⟶

Hyper Geometry Assessment Form

QUESTIONS: (Score 1-10)   **SCORE**

The stack has a minimum of 10 cards. _____

The "Next" button works on all cards. _____

The sounds are appropriate on all cards. _____

Rate the appearance of the stack. _____

The stack is easy to use. _____

The stack teaches basic geometry facts. _____

Rate the opening screen. _____

Total Score ⟶

Working With Charts And Graphs

CHARTING CHILDREN

In this lesson, students will make a number of graphs and charts using information that they collect about their classmates. They will interpret their graphs and charts, looking for trends and associations.

Grade Level: three to five
Duration: 60–90 minutes
Materials: Charting Children Data Collection Sheet (Pg. 83)
Software: *ClarisWorks*, Excel, *MSWorks*, *ClarisWorks for Kids,* or other spreadsheet/graphing program.
Intetnet Links: http://thegateway.org/index2/bargraphs.html

Procedure:

Before the Computer:

- Students will need to be familiar with those kinds of programs used for entering data and generating charts and graphs from that data. If they have not worked with these kinds of programs before, it would be a good idea to go through a sample lesson with them. Check your program's index or table of contents for the words "graphs" or "charts."

- Brainstorm methods of gathering the data so that it is done in an orderly and effective manner. Since the students will need to gather information on every student in the room, a system needs to be in place to ensure that this happens. Teachers should let the students develop the plan of action. Later they can evaluate their plan to judge how effective it was. For the first activity, *all* students will be gathering two kinds of information from *each* of the other students in the room, including themselves. The two pieces of information include the month in which they were born and their favorite color. The worksheet on page 83 should be reproduced for all students to use in their data collection.

- With the class, go over the directions on the Charting Children Data Collection Sheet. Discuss the section at the bottom of the worksheet that calls for a hypothesis. A hypothesis is a "guess" that a scientist tests many times to see if it is true or not. Students will be making a guess as to whether there is any relationship between the month in which they are born, and their favorite color.

- Have students gather data from their classmates using the Charting Children Data Collection Sheet.

- After all of the data is collected, and the graphs/charts are generated, they can see if there seems to be any connection. Students cannot really make a wrong hypothesis, but this test may show that this hypothesis is really not supported by the sample group of students in their classroom.

Working With Charts And Graphs

CHARTING CHILDREN (Cont.)

On The Computer:
- Using the Charting Children Data Collection Sheet, students make a number of graphs using the data. The first graph will show the distribution of favorite colors. A sample below was done using *Microsoft Excel* but students can use any spreadsheet program to generate graphs and charts. Their first step is to enter the names of the colors and then the number of students who said this was their favorite color. This has been done in the sample shown below.

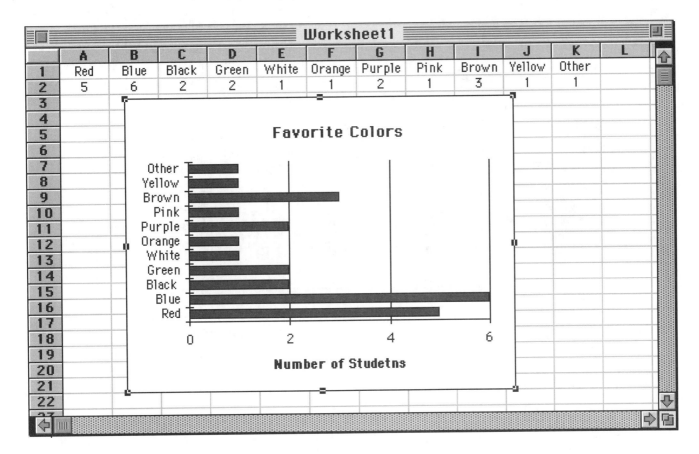

- The next step is to select the data. There may be several ways to do this depending on the program being used. Programs like *ClarisWorks*, *Excel*, and *MSWorks* also have an automatic charting menu or button. There are also many options so students can see how the chart is going to look before it is created.

- The chart shown here uses only one color and shows the distribution of all the color choices that were used in the survey. A title has also been added. Students should have enough information to create the first chart like the one shown above. They will notice that the total number of students in the survey above was 25. They can tell this by adding up the numbers for each color. After they have finished this chart, they are going to make another chart/graph for the distribution of birthdays in the class.

Working With Charts And Graphs

CHARTING CHILDREN (Cont.)

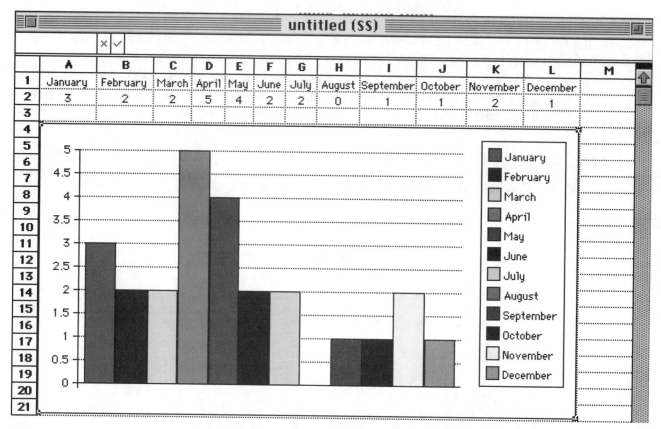

- In the chart above, which was done using *ClarisWorks*, a different color represents each month. The number along the left side of the chart shows how may students were born that month. For example, in the month of May there were exactly four students who said they were born during this month. Students should be aware that their chart will look different from this example, depending on the data that they have gathered.

- They will also notice that this chart has a color-coded key. This is shown along the right hand side. This key is very useful and similar to those found on maps. Students should have enough information to produce the second graph on the distribution of birthdays in their classroom. When they have finished with both graphs, they should print them for use in assessing their understanding of this activity.

- Students should print each graph and analyze their findings
    1. Is there a favorite color chosen most often? If so what is it?
    2. In what month did they find the greatest number of students born?

Assessment:

Students should refer to the original hypothesis. They should look at their data and graphs very carefully. Is there any kind of trend? For example, did they find that people born in May generally prefer blue over all other colors? They may find that there is no real connection between birth month and color preference. On a separate piece of paper, students should write their conclusions about the hypothesis and turn that in with their graphs. They should be very specific, using any evidence that was gathered during the activity.

Working With Charts And Graphs

CHARTING CHILDREN DATA SHEET

RULES:
- All students must be surveyed by everyone. How they do this is up to the class.
- Only **one** answer for each question is allowed. They can not say, "I like red and blue."
- Just record the month in which they were born, not the date.
- Students should not start working on the computer portion of this project until they have all of their data recorded.
- Each student does his or her own work. They should not allow someone else to do it for them.

Charting Children Data Collection Sheet

Name _____ Date _____

Student Name	Birth Month	Favorite Color
1.		
2.		
3.		
4.		
5.		
6.		
7.		
8.		
9.		
10.		
11.		
12.		
13.		
14.		
15.		
16.		
17.		
18.		
19.		
20.		
21.		
22.		
23.		
24.		
25.		

Colors: Red_____ Blue_____ Yellow_____ Green_____ Black_____ White_____
Orange_____ Purple_____ Pink_____ Brown_____ Other_____

Months: January_____ February_____ March_____ April_____ May_____ June_____
July_____ August_____ September_____ October_____ November_____ December_____

Hypothesis: Do you think there will be any relationship between the month a person is born and the color he or she prefers?
_____
_____

Ratio and Percent

WORKING WITH RATIO AND PERCENT

Students will be collecting data, analysing that data and using it in a presentation program to reinforce their understanding of ratio and percentage.

Grade Level: three to five
Duration: 60–90 minutes
Materials: computer, calculator, Data Collection Sheet (Pg.85), Practice With Percentages Work Sheet (Pg. 87)
Software: word processing program such as *Writing Center, ClarisWorks,* or *MSWorks, Hyperstudio* or other presentation program
Internet Links: http://yn.la.ca.us/cec/cecmath/math-elem.html

Procedure:

Before the Computer:

- Students should have some understanding of what the terms ratio and percentage mean. If this has not been discussed at all during the regular math curriculum, teachers should spend a little bit of time introducing these concepts.

- For the first activity, students will be doing activities that involve ratios. Teachers might want to explain that in some ways ratios are similar to fractions. A ratio compares two things. For example, students could create a ratio of boys to girls in the classroom. If there are 10 girls and 15 boys, a ratio of 10/15 exists between girls and boys in the classroom. If this ratio was true for the entire school, and students knew that there were 150 boys in the school, how may girls would there be?

$$\frac{10}{15} = \frac{?}{150}$$

- Page 85 contains a worksheet where students can gather data to create ratios. After they have collected the data for each item, they are going to use a word processor or drawing program to show some ratios both numerically and graphically. Most word processing programs are capable of importing images. They will need to be able to do this.

- Once students have established a ratio, they can project what the ratio would be as one of the numbers increases. While this projection is not really scientific it can still be useful.

- Teachers should have students begin gathering the data requested on the Data Sheet, page 85. When they are finished, they should look at the sample screen to put together their projects.

On the Computer:

- After students have completed the Data Collection Sheets, they should go into their word processing programs and generate some illustrated ratios such as the example, which was done with *The Writing Center*. A combination of clip art and text has been used to create a graphical illustration of a ratio, three books are being compared to four flags.

- Using the data students gathered, they should create a graphical illustration of all of the ratios. They will have a total of 10 screens that they can print, when they are finished. Follow the example as much as possible.

Ratio and Percent

WORKING WITH RATIO AND PERCENT

DATA COLLECTION FOR RATIO PROJECT

1. **Number of boys** _____ to **Number of girls** _____

2. **Number of brown eyes** _____ to **Number of blue eyes** _____

3. **Number of red shirts** _____ to **Number of white shirts** _____

4. **Number of sneakers** _____ to **Number of other shoes** _____

5. **Number wearing glasses** _____ to **Number not wearing glasses** _____

6. **Number blonde hair** _____ to **Number brown hair** _____

7. **Number of science books** _____ to **Number of books in desk** _____

8. **Number with short hair** _____ to **Number with long hair** _____

9. **Number with socks** _____ to **Number without socks** _____

10. **Number left-handed** _____ to **Number right-handed** _____

Directions:

- After you have completed the Data Collection Sheet, go into your word processing program and, using a combination of clip art and text, make some graphical illustrated ratios such as the picture shown here. Using the data you gathered.
- You will have a total of 10 screens that you should print when you are finished.
- Follow the example as much as possible.

3 BOOKS TO 4 FLAGS—A RATIO

Ratio and Percent

WORKING WITH RATIO AND PERCENT (Cont.)

Procedure: For Activity #2

Before the Computer:

- During this second activity, students will be working with percentages. Once again they will be using a word processing or drawing program to graphically illustrate their understanding of this concept.

- Teachers should review the concept of percent. Include the correlation between percent and decimals. Percentages are based on hundredths. A good example might be the batting averages in baseball.

- Review the use of calulators in finding percentages. Most computers have a calculator function or students could use personal calculators. Practice with the questions on page 87 and have students fill out the Practice With Percentages Work Sheet.

On the Computer:

- Having completed the previous two activities with ratio and percentage, students are now ready to put together a short presentation that shows their understanding of percentage.

- Have the students consider how they would teach someone about percentage who had never heard about the concept before. How would they go about doing it?

- Using either *Hyperstudio, ClarisWorks, Astound,* or some other kind of presentation program, students will be creating a short presentation that explains what percentage is and how it can be calculated. If, at this point, they do not understand about percentage, then they should review the previous activity or get some help from a class expert.

- On page 88, students will find a Precentage Lesson Pre-planning Sheet that will serve as an outline for how their presentations should look. They should be able to graphically illustrate the concept of percentage in **5 screens or less**.

Assessment and Follow-up:

Students should test the presentation on classmates and other students at the school. As a follow-up activity, have students look at the sports pages and figure out how the numbers for batting averages and team winning percentages are calculated. They will notice that they are expressed as a decimal rather than a percentage.

Ratio and Percent

PRACTICE WITH PERCENTAGES WORK SHEET

Sample Problem

- Joe Smith went up to the plate 100 times and he had 30 hits. Percent is usually expressed as a number between zero and infinity. We can determine Joe Smith's percentage of hits by dividing 100 into 30 and then multiplying by 100. Here is what it looks like using a calculator.

 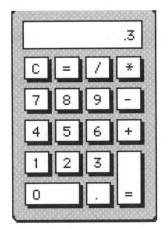 

- You should first enter the number 30. Then press the divide symbol. Now enter the number 100 and press the equals symbol. You should see .3 in the calculator window. For the last step, you should press the multiply button, enter 100, and then press the equals symbol again. You should now see a 30 in the calculator window. This stands for 30%. This means that Joe Smith gets a hit 30% of the time that he goes to the plate to bat.

Practice With Percentages:

- Using the sample problem above as a guideline, figure the answers to the following questions.

    1. On a test with 100 questions, Susan gets 47 correct. What percentage is correct? _____

    2. On a test with 25 questions, John gets 21 correct. What percentage is correct? _____

    3. Mickey goes to bat 76 times and gets 22 hits? What percentage are hits? _____

    4. There are 56 pieces of fruits in a bowl. Of these, 15 are bananas. What percentage are bananas? _____

    5. There are 123 cars at a car dealership for sale. Of these cars 27 are blue. What percentage of the cars at the dealership are blue? _____

Ratio and Percent

PERCENTAGE LESSON PRE-PLANNING SHEET

After you have completed this outline, you should put together a presentation that communicates the concept of percentage.

Title Card

Buttons/Links: _____

Notes (Text/Sounds/Animations): _____

Card 1

Buttons/Links: _____

Notes (Text/Sounds/Animations): _____

Card 2

Buttons/Links: _____

Notes (Text/Sounds/Animations): _____

Card 3

Buttons/Links: _____

Notes (Text/Sounds/Animations): _____

Card 4

Buttons/Links: _____

Notes (Text/Sounds/Animations): _____

Card 5

Buttons/Links: _____

Notes (Text/Sounds/Animations): _____

Math Journals and Dairies

KEEPING TRACK

Students will be keeping track of their math progress over a period of weeks or months. Their journal will become a record of what they have learned and the challenges that they faced in understanding the material

Grade Level: three to five
Duration: daily or weekly
Materials: computer
Software: *The Writing Center*, *ClarisWorks*, or a journal type of program, *Simple Text*.
Internet Link: http://www.qconline.com/www4kids/6-15-97/index.html

Procedure: For Activity #1

Before the Computer:

- Teachers should explain that the purpose of the diary/journal is to record experiences that are directly related to some aspect of the math curriculum. Depending on the computer schedule this might be only once a week or once a month. It could also be every day.
- Any good word processing program will work but there are a number of shareware titles that are designed to be journals. Teachers could download and try *My Journal Notebook 2.1* by MacLondon Creations and *Journal Keeper 2.0* by Randolph Rowan from the Internet.

On the Computer:

- When students are working on their math journals they should describe what they have learned during the week. If they can illustrate some of the concepts this would also be helpful.
- They will also want to make a note in their journals of those concepts that they are having trouble with or do not understand.
- *Please no private or confidential information should be recorded here.*

Assessment and Follow-up:

One of the ways to assess the math journal is to collect it either on a floppy disk or over a network. You may want to look for the following criteria when making an assessment.

1. Is there an entry for each computer session?
2. Is the content of the entry understandable?
3. Does the student share problems or concerns?
4. If possible, are graphics used to illustrate what is taught?
5. Assess the volume of writing in the journal; several paragraphs or just a line or two?
6. How is the spelling and punctuation?
7. Is the content consistent with what was discussed during those math periods?

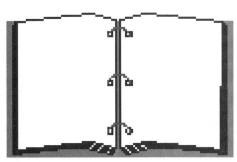

Math Journals and Dairies

TALKING JOURNALS

Procedure: For Activity #2:

On the Computer:

- Most newer computers come with some kind of voice-generated text-to-speech function. If students are working with a Macintosh, they can easily access this feature by using the program *Simple Text* that comes with the computer. *Simple Text* can also be used for their journal entries just as they normally would a word processor.

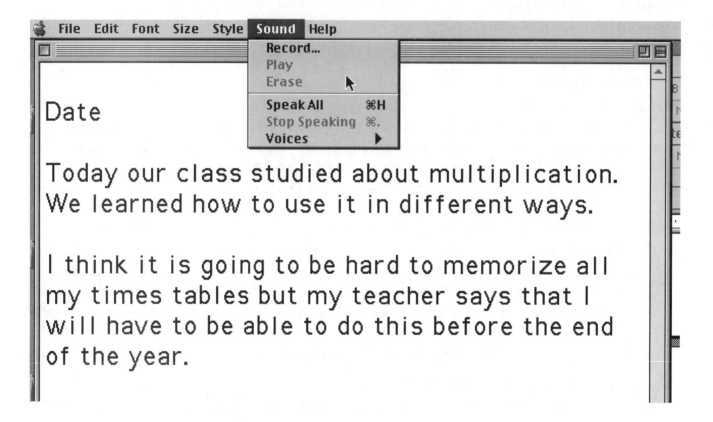

- To make their journal entry interesting and more exciting, students can add the speech function. Here is an example of an entry and the menu they will need to access in order to get the computer to speak the text.

- Enter text and move to the Sound menu. Students will find several different choices. One of these choices will make the computer read all the text that has been entered.

- There is also a record function at the Sound menu. Students can use this option to record a message that goes with this journal entry. Only one message is allowed per document and recording again will erase the earlier message.

Math Puzzles

MYSTERY SQUARES

Students will be creating their own mystery squares using a drawing program.

Grade Level: three to five
Duration: 30–60 minutes
Materials: computer, Mystery Cube Planning Sheet (Pg. 92), The Secret Code Planning Sheet (Pg. 94)
Software: Drawing program such as *Kid Pix*, *ClarisWorks*, etc.
Internet Link: http://www.oise.utoronto.ca/~wteo/humour.htm

Procedure: For Activity #1

Before the Computer:

- Students will need to be familiar with a drawing program and the tools available. If they have not worked with these programs before, teachers should go through a sample lesson with them.

- Depending on the age level, students may use addition or other math functions in their mystery cubes. After they create the cubes, they should print them out.

- Go over the example to the right with students. Notice that the sum of each pair of numbers is always 10. Notice that the sum of the three numbers moving horizontally always equals 15. This may look easy to do, but combinations take some thought. Students should use the Mystery Cube Planning Sheet on page 92 before going to the computers.

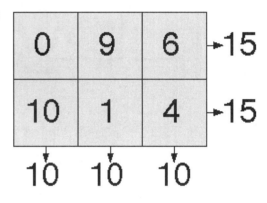

On the Computer:

- As they create the mystery cubes, they will need to make sure that they can solve them. The printed version **should not** include the answers, just the requirements such as, must add up to 10.

- The mystery squares can become more complicated when more squares are added. You may also require that both directions equal the same number as well as the diagonals. This is even more difficult.

A More Complex Example:

- Two new elements have now been added to the mystery square. First, there is a 3x3 layout with nine locations for numbers. Second, all rows and columns must add up to 15. Finally, only the numbers 1-9 are allowed and no number can be repeated. When they design their magic cube, they should leave the numbers out and print the empty cube with the desired totals.

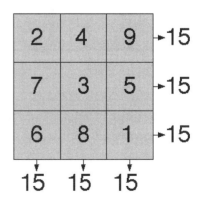

Math Puzzles

MYSTERY SQUARES
Mystery Cube Planning Sheet

Directions:
- Use the planning templates below to help you design some mystery squares. Notice that different shapes have been generated. The rules and results are left entirely up to you and your teacher. After completing the planning sheet, go to the computer and create your Mystery Cubes.

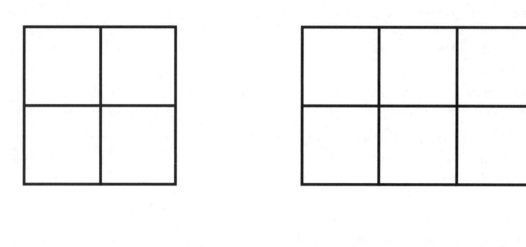

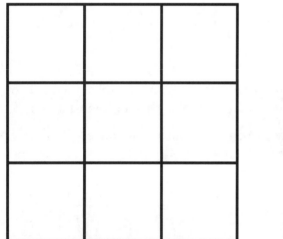

Math Puzzles

THE SECRET CODE

Procedure: For Activity #2:

Before the Computer:

- Discuss with students the concept and history of secret codes. A lesson might include the fact that during World War II most countries had their own secret codes so that they could pass information along to the troops without the enemy knowing what they were saying. Or have they ever wanted to be a detective and solve a mystery?

- In this activity, students are going to put together secret codes. After they have created their secret codes, they can write coded letters to other students in the classroom. After their first secret code, they may want to try codes that are even harder to decipher.

- A simple code to demonstrate, is this example where the letters have been inverted. If they want to use the letter A, then they would print the letter Z. As simple as this example is, if someone did not have the key to the code it would be difficult to determine what this message means. Can students figure out the coded message that is shown here?

- The code that they will be creating will be their own, and will be slightly more mathematical. There are many ways that they can come up with a formula for a code. Here are just a few ideas.

    1. Skip every other letter.

    2. Use two letters to stand for one.

    3. Use a combination of numbers and letters.

    4. Use only combinations of vowels to stand for all the letters of the alphabet (not easy to do).

    5. Count by 5's. (A=E, B=F)

    6. Use words to stand for letters (not easy to do).

MESSAGE

XLNKFGVIH   ZIV   UFM
_____   ___   ___

A=	Z
B=	Y
C=	X
D=	W
E=	V
F=	U
G=	T
H=	S
I=	R
J=	Q
K=	P
L=	O
M=	N
N=	M
O=	L
P=	K
Q=	J
R=	I
S=	H
T=	G
U=	F
V=	E
W=	D
X=	C
Y=	B
Z=	A

On The Computer:

- While students could easily write their codes, the purpose of this activity is to use the computer to generate not only the code but also their coded messages.

- The first thing they need to do is to plan out what their code will be for each letter. They do not want more than one letter to be represented by the same code. Therefore, each letter of the alphabet or number, if they choose to use them, should have a code equivalent. Use the Secret Code Planning Sheet to lay out the code and then enter it into their computer using a word processor or drawing program. Beneath the preplanning sheet they will also find a place where they can preplan some coded messages. Be very careful to write these messages in the proper code. **Always double check your work.**

© Teacher Created Materials, Inc.

Math Puzzles

THE SECRET CODE PLANNING SHEET

Alphabet																										
Code																										

Alphabet																										
Code																										

My Coded Message

Message 1: _

Translation: _____

Message 2: _

Translation: _____

Message 3: _

Translation: _____

Message 4: _

Translation: _____
_____

#2425 Integrating Technology Into the Math Curriculum © Teacher Created Materials, Inc.

Map Skills

MULTIMEDIA RESEARCH

During this activity, students will be working with multimedia reference CD-ROMs to do research related to map and globe skills. They will also create maps of a city that they invent.

Grade Level: three to five

Duration: 60–90 minutes

Materials: computer, Research Questions (Pg 96) Mapping The World Quiz Sheet (Pg. 97), Mapping My City Planning Sheet (Pg. 100)

Software: multimedia encyclopedia programs such as *The Visual Atlas, Groliers, Comptons,* etc., *Hyperstudio*

Internet Link: http://www.mapquest.com/

Procedure:

Before the Computer:

- Students will need to be familiar with multimedia type CD-ROMs and how to search for specific information relating to maps, countries, cities, and states. They will be using these to search for information and to use that information to complete the Research Questions on page 96.

On the Computer:

- During this activity, students will be asked to find various places around the world. They will also be asked questions about maps of the world.

- A sample question might be: What states border Arizona?

- To find the necessary information in *Grolier,* students should follow these steps:

    1. They would type in the word "atlas" or select it from the menu. Here they would be presented with a number of maps of places on the earth.

    2. The state of Arizona is listed as one of the maps that are available from this list. Enlarging this map, students can see that those states bordering Arizona are clearly marked, as shown in the sample.

    3. Looking at this map they can see that Nevada, California, Utah, Colorado, and New Mexico all share a border with Arizona. In addition the country of Mexico also shares a border with Arizona.

- When possible, they should print the area map.

© Teacher Created Materials, Inc.   95   #2425 Integrating Technology Into the Math Curriculum

Map Skills

MULTIMEDIA RESEARCH *(Cont.)*
RESEARCH QUESTIONS

1. Name the seven continents and the four major oceans. _____
   _____

2. On which continent would you find Spain? _____

3. What is the capital of Spain? _____

4. On which continent would you find Cambodia? _____

5. Name those countries that border Cambodia. _____
   _____

6. What famous volcano can be found in Indonesia? _____

7. On which continent would you find Iran? _____

8. On which continent would you find Angola? _____

9. How high is Mount Moko, which is found in Angola? _____

10. On which continent would you find Denmark? _____

11. Berlin is the capital of what country? _____

12. On which continent would you find Russia? _____

13. What is the capital city of Russia and about how many people live there today? _____
    _____

14. Ontario is a province (like a state) in which country? _____

15. Name the land and water boundaries of Ontario. _____
    _____

16. What are the names of the Great Lakes, which are found in the United States? _____
    _____

17. If you could travel anywhere in the world, where would you go? _____

18. Write a few sentences about the place you chose for question #17. _____
    _____
    _____

Map Skills

MAPPING THE WORLD

Procedure: For Activity #2:

Before the Computer:

- For the second activit,y students will be using *Hyperstudio* or any presentation program. For this example, *Hyperstudio* was used. Students should already be familiar with how to use this type of program.

- In order to do Activity 2, they will need a social studies book or multimedia CD-ROM that displays maps of the world.

- Depending on the grade level, teachers may choose to do a map activity for California, the United States, or the world. The example presented will be an activity covering the whole world.

On the Computer:

- Create a new stack in *Hyperstudio*. The first thing they will add to their blank card is a graphic image. *Hyperstudio* comes with a number of map images that are found in the HS Art folder. For this lesson choose the World Map. The one that students choose will depend on what they decide to map.

- Enlarge the map to cover as much of the computer screen as possible. To do this without changing the relative proportions of the map, make sure that the [G] tool is selected on the *Hyperstudio* tool bar. Click on the graphic and, while holding the SHIFT and APPLE keys at the same time, drag the map from the corner in (to make it smaller) or out (to make it larger).

- Students should have their social studies books or start the encyclopedia CD-ROM in the background that has maps on it. They should find a map that is close to the one that they selected for their card.

- There are several options for the next section of this activity. What they do will depend on the grade level and the map they have chosen.

- Here is a sample screen from *Hyperstudio*. A number of activities can be done with this map.

- Create invisible buttons (check your *Hyperstudio* manual on how to do this) that identify all the continents and the major oceans. These same buttons can report vital information about each continent and ocean. For example: the number of countries, estimated population, number of square miles, important agricultural products grown, religions practiced. All of this information can be obtained from a multimedia reference disk.

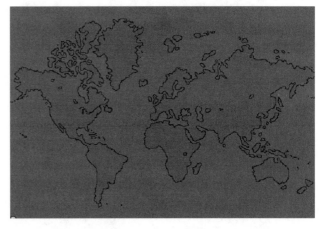

- If a map of the United States is used, similar types of activities can be done. Names of the states, population, state capitals, types of industries, and weather/climate can be found.

- If a map of California is used, the information can include the counties found in California, the four main regions, state capital, agriculture and industry, and Native Americans and where they live(d).

Map Skills

MAPPING THE WORLD

Quiz Sheet

It's Time For A Quiz! (for those students who did the world map) Label the continents and the oceans on the map below.

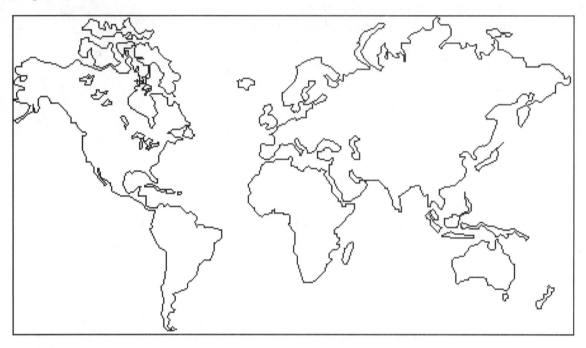

It's Time For A Quiz! (for those students who did the U.S.A. map) Label as many states as you can. Use the abbreviations when necessary.

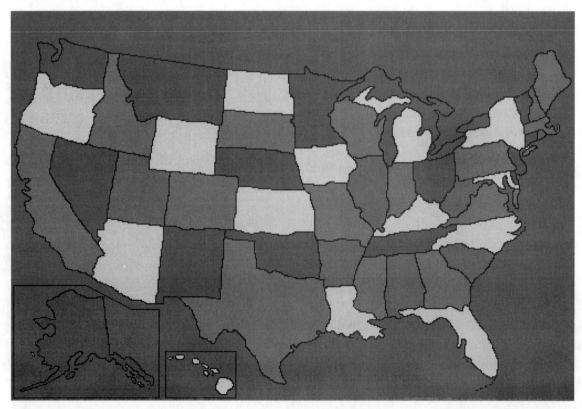

Map Skills

MAPPING MY CITY

Procedure: For Activity #3

Before the Computer:

- During this activity students will be using a drawing program to create several different kinds of maps.
- Discuss with them the history of map making. A person who creates maps is called a cartographer. In order to make a map that is an accurate representation of a land mass, a considerable amount of measurement must be done. Early maps of the world and America were not very close to what these land masses actually are. Today we can map the globe from outer space and generate maps of the oceans and land masses that are quite accurate.
- Maps are used every day to help people find their way from one location to another. All major cities and towns have maps that show the streets, city offices, parks, and other important landmarks.
- Most local maps will have a compass which shows direction and also they will use a grid so that when one looks up a street, it can be found at the intersection of two points on that grid.
- Students should be familiar with a variety of maps. They should be able to determine the direction, the grid location of streets, and important landmarks. In addition some maps will have a **map key** that describes what the symbols used on the map represent. They should be able to read this key.
- On page 100, there is a sample plan for a city called "My City." Students will be making an outline for their very own city. After they have studied the example, they can draw up a plan for their town. When the preplanning sheet is finished then they can move over to the computer and start creating their maps.

On the Computer:

- The next page contains a partial drawing of "My City." This was created using *ClarisWorks*, but any drawing program will work. Students should complete the preplanning sheet before they actually start drawing.
- The best place to start with the map is to create the grid. On "My City" they will see numbers running down the left-hand side and letters across the top. This makes it easy for someone reading this map to determine where a street should be located.
- They should notice that the compass rose tells the direction.
- The city needs to include a compass rose, streets, parks, city hall, schools, a river, a library, a police station, and other important buildings or landmarks.

Map Skills

MAPPING MY CITY

Directions:

- Study the sample city. Use the Preplanning Sheet to create your own city. Be sure to include the grid locators along the side of your map. Your city needs to include the following:

    a. compass rose b. streets c. parks

    d. city hall e. schools f. a river

    g. a library h. a police station i. other important buildings or landmarks.

- You can use any names that you wish for the streets and other items. Use your imagination but be sure to include **all of** the items.

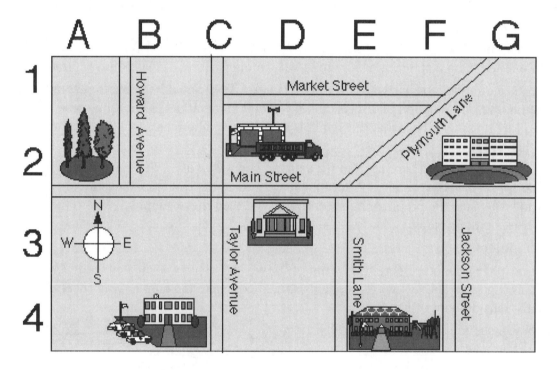

Preplanning Sheet For "My City"

#2425 Integrating Technology Into the Math Curriculum   100   © Teacher Created Materials, Inc.

Map Skills

MAPPING MY CITY (Cont.)

Assessment and Follow-up Activity:

Once you have created a map for "My City," you are ready to generate some questions regarding your maps. For example, look at the sample map on page 100 for a moment and see if you can answer these questions.

1. At what grid location would you find a park? Remember a grid location consists of a number and a letter. B3 would be an example of a grid location but it is not the answer to this question.

   _____

   _____

2. From Taylor Avenue what direction would you travel to get to Jackson Street and which road would you use to get there?

   _____

   _____

3. From the Hospital, what direction would you travel in order to get to the Fire Department?

   _____

   _____

4. The school is located between what two streets or lanes?

   _____

   _____

5. What grid location would you give for the Police Department, which is located west of Taylor Avenue?

   _____

   _____

Now it is your turn to make up some questions for the map that you created. Use the space below for that purpose.

Finding Math In Sports

SUPER SPORTS LEAGUE

During this activity, students will be investigating the correlation between math and sports. They will be inventing their own sports teams and doing calculations on win/loss percentages. They will become statisticians and keep statistics on players.

Grade Level: three to five
Duration: 60–90 minutes
Materials: computer, newspaper, Super Sports League Questionnaire (Pg. 104), Running Back Statistic Form (Pg.106)
Software: word processing program and a computer calculator
Internet Link: http://www.adsnet.com/conan.brown/

Procedure: For Activity #1
Before the Computer:

- Students should look in the sports page of their local newspaper. Depending on the time of year, they should find a page that lists professional sports such as baseball, basketball, football, and hockey.

- Teachers should discuss with students the team standings and win/loss records and explain how numbers such as "Games Behind" and "Winning Percentage" are calculated.

- During this activity they will be inventing their own sports league and team standing charts. There are four things that they will be concerned with when creating the team standing charts. These four are: wins, losses, percentage, and games back, or abbreviate these to: W, L, Pct., GB.

- Students should pick any sport that they like, but there are some **rules** they must follow.
    1. The league must have **10 teams** which are divided into two conferences, like the Eastern Conference and the Western Conference.
    2. The teams in the league will have already played an imaginary 80 games, no matter what sport is chosen. Make sure the win/loss record adds up to 80 games.
    3. No two teams can have the same win/loss record.
    4. Students can name the teams anything that they like, and give them any win/loss record.

- Once they have created their **10 teams which are divided into two conferences**, they will be using a calculator to help determine mathematical information about the teams.

- Students may design a team "Logo" using a drawing program.

- On page 103, students will find a sample of this activity. The sample will be used to explain how to find the mathematical information that they are looking for so that they can fill in the information for the Percentage and the Games Back headings.

Finding Math In Sports

SUPER SPORTS LEAGUE (Cont.)

WEST

TEAM	W	L	Pct	GB
Jayhawks	67	13	.837	—
Bluefish	62	18	.775	5
Yellowbacks	59	21	.737	8
Dinos	44	36	.555	23
Demons	23	57	.287	44

EAST

TEAM	W	L	Pct	GB
Hotdogs	69	11	.863	—
Angels	64	16	.800	5
Redtails	50	30	.625	19
Snakes	34	46	.425	35
Nerds	14	66	.175	55

Before the Computer: *(Cont.)*

- The chart above presents an example of what students will be doing. Notice that there are **two conferences**, the West and the East. Also notice that there are 5 teams in each conference and that each of these teams has played a total of 80 games.

- In order to find the Percentage (Pct), students should take the games won and divide it into the total number of games. Using the "Jayhawks" as an example, divide 67 into 80 on a calculator. The result would be **.837**. This represents that team's winning percentage.

- To find the "Games Back" number, add up the difference between the games won and the games lost, then compare that to the first place team. In the Eastern Conference for example, the Hotdogs had the best record. In the "Games Back" column, they will see "—." This means that no other team had a better record. The second place team was the Angels. They were 5 games back in the win column and 5 games back in the loss column. 5+5=10. Divide 10 by 2 to find that they are 5 games behind the leader. When looking at the standings in a newspaper, it is very likely that **not all teams will have played the same number of games**. This is not true of the teams that students will be creating. All they really need to do is find the difference between the games won or the games lost, in relationship to the leader. Compare either one, and they will come up with the correct answer.

On the Computer:

- The students should take their own sports league that they created and make a chart similiar to the sample. Remember to follow the rules as outlined on page 102.

- They may use a drawing program or other word processing program. After they have finished creating the league standing chart, start doing the math needed to determine the percentage and games back. They can use the computer calculator or a personal calculator.

- When the work is complete and they have printed their charts, they are going to generate some questions about their teams, the league, and the conferences using the Super Sports League Questionnaire on page 104.

Finding Math In Sports

SUPER SPORTS LEAGUE QUESTIONNAIRE

Assessment Questions:

Create 9 questions regarding your sports league that other students could answer from your league standings chart. Exchange a copy of your chart and these questions with another student. Work independently and answer each other's questions.

1. _____
   _____

2. _____
   _____

3. _____
   _____

4. _____
   _____

5. _____
   _____

6. _____
   _____

7. _____
   _____

8. _____
   _____

9. _____
   _____

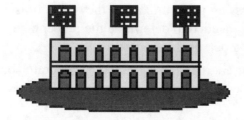
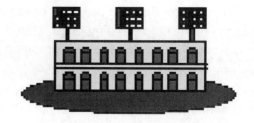

Finding Math In Sports

WATCH RANDY RUN

Procedure: For Activity #2:

Before the Computer:

- During this activity students are going to become "statisticians." This is a person who keeps statistics on players. Managers and coaches rely on these statistics to make changes in the way their teams are playing. People are actually paid large sums of money to sit behind a computer and calculate this statistical data and then put it into video format so that it can be "superimposed" while the action of the game continues.

- It is important that the student have a basic understanding of the game of football as it is played in the United States. The only player on the football team that they will be concerned with right now is the one called running back. This is a player who is handed the ball so that he can run it forward.

- Review how to find the average of numbers, and students will also need an understanding of how *ClarisWorks* or other word processing programs generate graphs. On page 107, students will find an outline/worksheet for this activity.

- Before they get on the computer, students need to create a list of ten football players who are running backs. By the way, girls can make their running backs women, even though there is no professional football league for women, yet. They can give them any names that they like. Second, they will need to **invent** some statistics for them. This will not be hard but they will need to follow some rules.

- Usually a coach is looking for two things in a running back. These two things are: the number of times the running back carries the ball and the total number of yards that the back gains. For example: Joe Smith might carry the ball 10 times and gain 30 yards. This is not very good because it means that Joe is only making 3 yards a carry. On the other hand, if Joe Smith carried the ball 10 times and gained 150 yards, this would be very good because it means that he is gaining 15 yards per carry.

- Students' running backs can carry the ball as many times as they want them to, but they **must gain yards. The number of yards gained must be between 1-200 for each game played.**

Finding Math In Sports

WATCH RANDY RUN *(Cont.)*

Running Back Statistic Form

- Here is an example of a partial list of running backs and their statistics for each game. You will be using the form below to create your list of running backs. You will also need to use the computer calculator to do some math for their statistics.

RB	Game 1 C/Yrds	Game 2 C/Yrds	Game 3 C/Yrds	Game 4 C/Yrds	Game 5 C/Yrds	Game 6 C/Yrds	Game 7 C/Yrds	Game 8 C/Yrds	Game 9 C/Yrds	Game 10 C/Yrds
Smith	12/99	7/78	16/132	10/102	5/56	18/122	5/45	19/145	7/66	23/178
Jones										
James										
Howard										
Rosten										

- Part of this chart has been filled out. Smith, who is the first running back listed, has his statistics posted for the first ten games. The first number shows how many times Smith carried the football in game number one. The second number shows how many yards Smith gained. Later you are going to add up all the yards for games one through ten and then divide this number by ten, the number of games. This will tell you what Smith's average yards per game was. You are also going to add up the number of times that Smith carried the ball and divide this into the total number of yards he gained. This number will tell you what Smith's average gain per carry was over the ten games played. Later you will be making charts and graphs of this information.

- Use the chart below. Create 10 running backs and their statistics for 10 games.

RB	Game 1 C/Yrds	Game 2 C/Yrds	Game 3 C/Yrds	Game 4 C/Yrds	Game 5 C/Yrds	Game 6 C/Yrds	Game 7 C/Yrds	Game 8 C/Yrds	Game 9 C/Yrds	Game 10 C/Yrds	Total C/Yds
1. _____											
2. _____											
3. _____											
4. _____											
5. _____											
6. _____											
7. _____											
8. _____											
9. _____											
10. _____											

Finding Math In Sports

WATCH RANDY RUN *(Cont.)*

On the Computer:

- Students should have 10 running backs, each with statistics for 10 games played. The last column on their worksheets is to be used to find the total number of carries and the total number of yards. Record these numbers for each of the 10 running backs.
- They are ready to create some graphs of the statistics that they have developed. They need a program like *ClarisWorks, MSWorks, Excel,* or *File MakerPro*. These are programs that will allow them to make spreadsheets and instantly make a graph of the data. Have students review the graphing section of the program with which they will be working.
- The following example of a spreadsheet was made using *ClarisWorks*. Please note that the **average** yards per game was used in the chart. In their totals column of the worksheets, they already have added up all the yards from the ten games played. To find the **average per game**, just divide this number by 10.
- The graph below shows all ten players, their total number of carries, and the average yards per game that they gained. The data entered into the spreadsheet, was converted into the graph using *ClarisWorks* also. To create the graph, all the data was selected using the Edit-Select All menu. Then Make Chart was selected from the Options menu. Please note that Make Chart will not appear until some data is selected.

A	B	C	D	E	F	G
Player	Carries	Average YPG				
Smith	155	76				
Jones	189	106				
Williams	90	99				
Rosten	123	44				
Hayes	101	23				
Unger	201	112				
Rolf	66	77				
Sanders	144	89				
Walker	178	56				
Phillips	232	103				

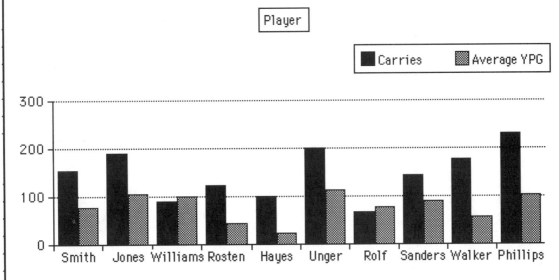

© Teacher Created Materials, Inc.   107   #2425 Integrating Technology Into the Math Curriculum

Finding Math In Sports

WATCH RANDY RUN (Cont.)

Assessment :

- Even though the data for this activity was made up by the students, it should still be possible to ask intelligent questions about the data. As long as they followed the rules when setting up their worksheets, they should not have a problem with this section. For the assessment of this activity, students should create 5 questions that they could ask a classmate about their data, spreadsheet, and graph. The following is an example of some questions to ask using the sample on page 107. In your opinion, who was the most **efficient** runner in the sample? (HINT- There are only two runners who actually have **less** total carries than yards per game earned.) In your opinion, which runner(s) would you trade right away because of their poor performance? Their questions should be interesting and unique. Students might also want to compare results with some other students to see who has the best or most efficient running back in the classroom. They could also find out who has the worst or least productive running back.

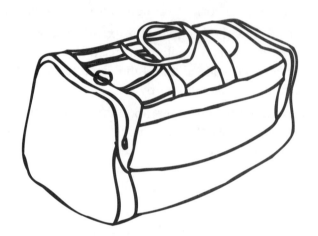

Follow-up Activities:

- The kinds of things that were done during this activity, would also work for many other sports. In baseball, for example, pitchers are given an earned run average, also called an ERA. This number is determined by how many innings they pitch and how many earned runs they allow during those innings.

- In basketball, players have statistics kept on every aspect of their game. Three point shooters are given a percentage that is computed by dividing the total number of three point attempts by the number of three pointers made. The very best shooters make between 40-50% of their three point shots.

- How could students use what they have learned to do a statistical analysis of other sports?

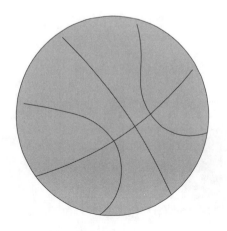
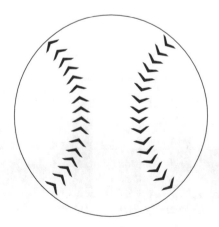
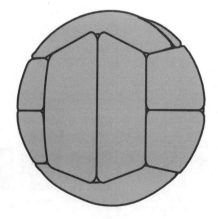

As Simple As Giving Directions

DUPLICATE ME

During this activity, students will learn how to give clear and concise directions to complete several defined tasks.

Grade Level: three to five
Duration: 60–90 minutes
Materials: computer, Duplicate Me Shapes (Pg. 110), Robot Planning Sheet (Pg.112), The Pet Expert Sheet (Pg.113)
Software: word processing program with drawing capabilities
Internet Links: http://www.geom.umn.edu/apps/

Procedure: For Activity #1
Before the Computer:
- Explain to students the importance of following directions. If a single step is left out when following a recipe, for example, what they are making will probably not turn out properly. Many products are sold unassembled, which means when it is brought home from the store, they must be able to put it together. Skipping steps or trying to put the purchased product together without following directions can spell disaster.
- During this activity, students will be **giving** directions to other students. They will be using these directions to complete a task. Each individual will know what his or her finished product is supposed to look like, but noboby else will.
- As an example of something that they might do in this activity, look at the shape below. How would students give a classmate directions to recreate that shape if they could not see it in advance?
- They might give these directions:
    a. You are going to draw a geometric shape.
    b. This shape has opposite sides that are equal in length and parallel to each other.
    c. First draw a line one inch long (2.54 cm) vertically in the middle of your paper. (Vertically means north to south.)
    d. From either end of the first line that you drew, draw another line 3 inches (7.62 cm) horizontally and at a right angle to the first line.
    e. The third line will look just like the first line. It will be the same length (1 inch or 2.54 cm) and it will be connected to the end of the second line.
    f. The fourth line will be the same as the second (3 inches or 7.62 cm) and it will join all the lines together so that they form a common geometric shape.
- It is hard to believe that it took that much in the way of directions to explain how to draw the shape above. Even with all the directions, it is possible someone could get it wrong if they did not know the meaning of words like, parallell, vertical, horizontal, and right angle. Students should see that giving accurate directions is a lot more difficult than it looks.

© Teacher Created Materials, Inc. 109 #2425 Integrating Technology Into the Math Curriculum

As Simple As Giving Directions

DUPLICATE ME

On the Computer:

- For the first activity in this section, students are going to pick from one of the shapes shown below and write **all** of the directions they think a person would need to be able to recreate that shape.
- Once they have all their directions together, they can print them and give them to classmates to recreate just by following their directions. Students will need a copy of this page with the shapes on it, so that they can show others what the shape was **supposed** to look like.

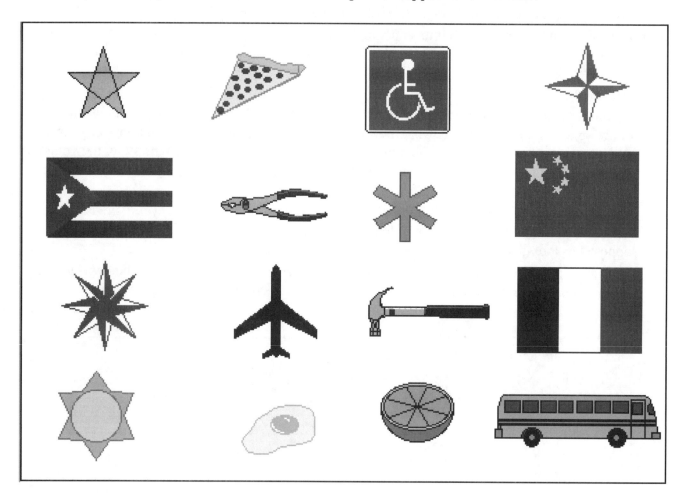

Student Directions:

- Pick one image at a time. Do not show the other person the image that you selected, or tell them in advance. Do not say for example, "It's an airplane." Describe what the object looks like, giving clear directions of how to draw it.
- Give the description to another student and have them draw what they think the image is, based on the description that you write down. They must use the computer to draw.
- Taking turns, you can do as many of the shapes as you wish, or make up a shape of your own and write a description for it.

As Simple As Giving Directions

MY VERY OWN ROBOT

Procedure: For Activity #2:

Before the Computer:

- Have students imagine what it would be like to have their very own robots. They could make them do all kinds of things and go to all kinds of places.

- Explain that if they did own robots, they would probably be controlled by a computer of some type. The problem with robots, unlike people, is that they usually will do exactly what students command them to do. If they say turn left, the robot would turn left. If they say jump up and down ten times, this is exactly what the robot would do if it could.

- In this activity, students are going to get to control their very own robot. The robot is going to be one of their classmates. Students are going to write their robot a set of instructions on the computer, and the robot is going to follow these instructions.

- The first task will be simple so that they can get used to giving directions that are exact. Students will describe what needs to be done to get from their desks to one of the doors of the classroom. They **cannot** simply say, "Stand up, walk to the front door of the classroom and open the door." The robot must know direction and distance. Here is an example of a classroom.

- In this example the student is marked with an "X." The positions North, South, East, and West are all relative to where the robot is standing. Here is what directions might look like in this classroom for student robot "X."

1. Robot stand up.
2. Robot turn west.
3. Robot take 5 steps forward and stop.
4. Robot turn south or north.
5. Robot take 12 steps forward and stop.
6. Robot turn west.
7. Robot take 9 steps forward and stop.
8. Robot open door and walk out of the room.

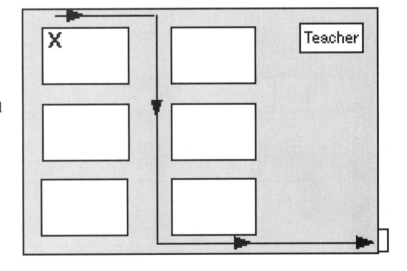

On the Computer:

Okay, it is the student's turn. He or she should write a clear set of directions to move from his or her desk to the door of the classroom, and then out of the room. Depending on where his or her desk is located in the classroom, he or she may have more or less instructions than in the sample above.

Assessment:

A true assessment of this activity is the achievment of the goal of getting to the classroom door and then out. Is the student robot able to understand and execute the directions as they have been written? Are changes needed in the directions? If so make them and try again.

As Simple As Giving Directions

THE WAY TO SCHOOL

Procedure: For Activity #3:

Before the Computer:

- During the last activity, students gave relatively simple directions for their student robots to move from their desks and out the door of the classroom. In this next activity, they will be giving sets of directions to do more complicated tasks.

- The first task will be for the students to describe in detail how a robot would get from their house to their school. Use the form below to complete all the directions necessary for this task. If they do not know the names of streets, landmarks can be used, for example restaurants, buildings, trees, etc.

On the Computer:

- After they have written their sets of instructions on the computers, they will also need to create maps of what they think the path is from their homes to the school. In order to do this they will need a drawing program and they will also need to be able to visualize how the robot moves. This is not easy to do.

Planning Sheet (Get your robot from your home to school.)

Step 1: _____

Step 2: _____

Step 3: _____

Step 4: _____

Step 5: _____

Step 6: _____

Step 7: _____

Step 8: _____

Step 9: _____

Step 10: _____

Step 11: _____

Remember that you must also draw a map of what you think going from your home to school looks like. Do not show your map to anyone but your partner. When you are finished with your set of instructions, you will want to give it to other students in the room to see if they can find out where you live.

As Simple As Giving Directions

PET EXPERT

Procedure: For Activity #4:

Before the Computer:

- Teachers should discuss with students the idea of owning and caring for a pet. Most students probably either have a pet or know someone who has a pet. There are all kinds of pets from cats, to dogs, fish, turtles, lizards, birds, mice, rats, hamster, etc. Each kind of pet requires special care. If they have ever had pets, they know first-hand that they will not live unless special care is taken. With pets, it is very important that directions are followed. These directions are usually given by an expert, someone who is knowledgeable about their kinds of pets.
- In this activity they are going to be the experts. They should choose pets that they know a lot about or that they have. They should not give instructions about how to care for and feed lizards if they have never had one before.
- In the exercise below, they will be asked to give specific information about the pets that they have chosen. They should fill in The Pet Expert forms. What they will be doing is creating a database of information about their particular creatures so that others can use it. There are no **wrong** answers to the questions below and please *do not use an encyclopedia* to get the information. Remember students are supposed to be the experts.

On the Computer:

- After students have filled in the information needed for each question, they will be putting all of their data into paragraph form. This will allow anyone who wants expert knowledge about a particular pet to use their data.

The Pet Expert

Kind of Pet:_____ Your Name:_____

1. What kind of things should a person look for when selecting this kind of pet?

    a. _____
    b. _____
    c. _____

2. What does a pet like this cost?

3. What kind of foods does this pet like to eat?

    a. _____
    b. _____

4. What kind of shelter does the pet need and how much does it cost?

    a. _____
    b. _____

5. Are there any special problems or concerns that a person needs to know before buying a pet like this?_____

6. On average, how long does this pet live? _____

7. Why would someone want this kind of pet? _____

As Simple As Following Directions

REALLY READING!

During this activity, students will learn how to follow clear and concise directions to complete a defined task.

Grade Level: three to five
Duration: 60–90 minutes
Materials: computer, Directions for Problem 1, 2, 3, 4, Self Assessment Sheet (Pg.115)
Software: word processing program with drawing capabilities
Internet Link: http://www.groton.k12.ct.us/WWW/cb/math.html

Procedure:

Before the Computer:

- This activity will focus on the students' ability to follow a number of commands. The finished product will be a geometric shape. Students must use a computer drawing program, preferably one that has a ruler available.
- During the last activity, students had some practice with giving directions, now they will find out how well they can follow them.
- The answers to each of these sets of directions can be found on page 141. Students should not peek until they have first completed each part of this activity.

On the Computer:

- **Problem #1 Directions:**
    1. Draw a square with sides equal to 2 inches.
    2. Inside this square, draw a circle that takes up most of the inside space of the square.
    3. Make the circle red.
    4. Put a large X inside of the circle.
- Students should print their geometric shape and save it for later so that they can compare it to the answer on page 141.
- **Problem #2 Directions:**
    1. Draw 3 circles each with a diameter of 1 inch.
    2. Put these circles side by side horizontally so that they just touch each other and flow from left to right in a straight line.
    3. Color the first circle blue, the second circle green, and the third circle yellow.
    4. Draw a rectangle around all three circles that just touches the outside edges of the circles.
    5. "Send" the rectangle to the "Back" so that is does not cover up your circles but simply creates a frame.
- Students should print their geometric shape and save it for later so that they can compare it to the answer on page 141.

As Simple As Following Directions

REALLY READING!

On the Computer: *(Cont.)*

- Although they may not think so, the first two problems on page 114 were relatively simple. The difficulty with directions is that they must be written in such a way that most people will be able to understand them and follow through with the steps.
- The next set of problems will be more complex and involve more steps. They will have to decide for themselves what the directions mean.
- **Problem #3 Directions:**
    1. Draw a triangle with a base of about 2 inches and equal sides of about 2 1/2 inches. Most drawing programs will have a polygon tool.
    2. Color this triangle green.
    3. Draw 3 circles each about 1 inch in diameter. Place each circle on the outside "points" or vertices of the triangle. Pretend that you are trying to balance these circles on these "points."
    4. Fill the circles with black, red, and blue. You decide which color to give each one.
    5. Draw 3 lines that connect the outside edges of each circle to the other. Each circle will be touched by 2 lines and all the circles will be connected.
    6. Type your name and place this at the bottom of the triangle.
- Students should print their geometric shape and save it for later so that they can compare it to the answer on page 141.

Self-Assessment:

Please answer the following questions about problems 1, 2, and 3.

1. What did you find was the most difficult part of this activity? _____

2. If you have compared your drawings to the answers found on page 141, were any of your drawings close to the answers? If yes, why do you think the drawing(s) came out correctly?

3. If your answers were far off from the actual answer, what do you think happened?

4. Is is possible that the instructions were just "lousy?" _____

5. How do you think you could have made them better? _____

 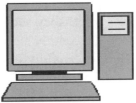

As Simple As Following Directions

CHECK OUT THESE CHECKERS

Problem #4:

- For this final activity, students will need to use a program that will allow them to draw shapes and then move them around the screen. Programs that do this are *ClarisWorks* or *MSWorks*.
- In this activity, they will be following all the steps needed to create a playable checkers game. This will be a two player game.

On the Computer:

- If students are using *ClarisWorks*, they will be selecting Drawing from the opening screen.
- The game of checkers uses an 8 by 8 checkerboard playing surface. The colors alternate between light and dark and the "men" can only move on the dark squares.
- If possible, turn the rulers on so that they are dividing the screen into "centimeters." Normally, they will be set for inches so students will need to find the menu that allows them to set up the rulers differently. In *ClarisWorks*, they will find this function in the Format menu.
- Draw a square near the top left corner of the screen that is 1 centimeter on each side.
- Color this square brown. In *ClarisWorks*, objects are colorized by first clicking once on them and selecting the color from the color option in the Tool palette. If they do not see the tools, they can turn them on by going to the View menu and choosing Show Tools.
- While the first object is still selected (there are handles around it) go to the Edit menu and select Copy.
- Since students need a total of 64 squares and 32 of them are going to be brown, they are going to Paste 31 more of these squares over the first one that they drew. They can either select Paste 31 times from the Edit menu or use the shortcut of Apple+V. It will appear that nothing is happening but if they copied the first shape correctly the new shapes will be pasted over the top of the first one.
- If students are using *ClarisWorks*, make sure that the "autogrid" function is on. They should see centimeter lines going both directions.
- In order to move a square into the proper location, click once on it and use the up, down, left, right, arrow keys.
- It would be a good idea to draw a square that covers the entire 8x8 area of the checkerboard and then send it "to the back."
- Once they have placed the 32 brown pieces in the correct locations, they will be ready to start on the yellow pieces.
- An example of where they should be at this point, is on page 117. Example #1 shows the 32 brown squares, the centimeter rulers, and a box that encloses the entire playing surface. Be sure students study this carefully if they are not that familiar with the game of checkers.

As Simple As Following Directions

CHECK OUT THESE CHECKERS (Cont.)

On the Computer: *(Cont.)*

Example #1:

- For the next part of this project, they are going to put 32 yellow squares between the brown ones.
- Select by clicking once on the brown square and choose Copy from the Edit menu. Now choose Paste and move this new square over to a blank space.
- Change the color of the new square to yellow.
- Next, they are going to Copy this square and Paste it 31 more times. They can begin to move them around using the arrows keys. There are 32 locations in which to place the new yellow squares. When they are finished, their checkerboards should look like this. **If they are not happy with the spacing, turn off the autogrid and they will be able to make a finer spacing.**

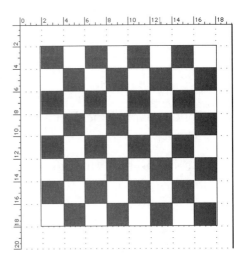

Example #2:

- Example two shows the completed game board. All that is left to be done is to create the game pieces and talk about how to play this game with a partner.
- If student drawings do not resemble this one, go back and recheck the steps to see if they might have made a mistake. If they are still stuck, they can talk to a classroom expert and see if he or she can help.
- *The geometric arrangement of the playing surface is very important.* If they do not have a total of 64 squares the game will not play correctly. Also, the squares must alternate as in the pattern shown here. The choice of color is really up to them. Yellow and brown were chosen because the next set of directions ask students to create red and black circles to use as the playing pieces. These will be the objects that they will move around the playing surface. Although the squares are still movable, they will not be moving them during a game.

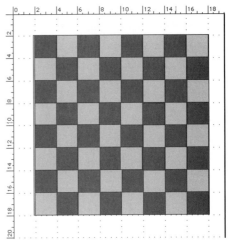

- At this point, be sure students save their hard work. Use a filename like "Checkers."
- Now students are ready to make the game pieces. The game pieces will consist of red and black circles. When a game piece is ready to be "kinged," all they will need to do is change the color to identify it as a king. The color blue would work well for the black player and the color green for the red player.
- Select the circle drawing tool and make sure that it is set for "filled" mode. Draw a circle with a diameter of around 3/4 centimeter. This means that the circles (game pieces) will cover up most of the brown squares that they occupy.
- Each player will have a total of 12 pieces. Start with either the red or the black using the same copy and paste technique that was used earlier to create the checkerboard squares.

© Teacher Created Materials, Inc.

As Simple As Following Directions

CHECK OUT THESE CHECKERS (Cont.)

Example #3:

- When all 24 circles (games pieces) have been created, assemble them on the playing board as in Example #3.

- The finished playing board should look like the example if they have followed all of the directions for this activity.

- Be sure to have students **Save** their work.

- If they are playing a game and want to quit before the game is over, make sure that they save their game to a different name other than the start-up game board. This way they will not have to reset all of the playing pieces each time they start a new game.

- If they are not sure about the rules for the game of checkers, they will need to ask an expert or do some research to find out what the rules are.

- The game they have created is for **two players**. Decide who will go first.

- In order to move the pieces around the game board, first select the pointer tool from the tool bar.

- Students should click on the game piece that they would like to move and then move it.

- When they jump an opponent's man, the simplest way to remove his game piece is to click once on the piece and then press the delete key.

- If for any reason they accidentally move one of the background pieces, they can immediately go to Edit on the menu bar and select Undo.

- If they want to prevent the game board from ever moving, they can use the Edit menu, Select All, and then in the Arrange menu choose Lock.

- **This must be done before they place the game pieces down however, or they will lock the game pieces from movement as well.**

- In all other respects, this game will play just like a standard checkers game.

As Simple As Following Directions

CHECK OUT THESE CHECKERS *(Cont.)*

Making Kings:

- While they could have created individual King pieces, there is a simpler solution to turning a piece into a king once it reaches the other side of the board. Suppose students have been playing for a while and their board looks something like the board shown below. Some of the black pieces are very close to becoming kings. To make them kings, students should decide first on colors for the kings. In this example, the black pieces could have blue kings and the red pieces could have green kings.

- First the pointer tool is selected. Then the piece that they wish to "king" is selected. Next the color palette is selected and the new color of green is chosen.
- This piece is now ready to be played as a king which means that it can move backwards and forwards and jump in either direction.
- Remember to save the game to a different name if they are not finished playing. Opening up the original file, which was saved as Checkers will always present them with a board and pieces that are ready to start a new game.

Assessment and Follow-up:

This activity was by far the most difficult of the series of activities in this unit. If students were able to get a finished product that works (as in the examples above) they deserve a round of applause.

Now that they have some experience in following directions to make a game, they could try using the same game board, but create a different game. It turns out that the game of chess uses the same size playing board. They will need to draw shapes that represent the game pieces found in chess, which is a little more difficult. They will also need to know the rules that apply to the game of chess. With some patience on their part they should be able to create a very attractive and playable game of chess.

Activities Using Logic

IF-THEN STATEMENTS

Students will be working with logic during the following activities. "If-Then" statements will be used to show how logical decisions can be made by the computer.

Grade Level: three to five
Duration: 60–90 minutes
Materials: computer
Software: *Hyperstudio*
Internet Link: http://www.allaboutgames.com/robot/fetch!9862587811

Procedure:

Before the Computer:

- Many students may already be using logic without even knowing it. In a way, logic is nothing more than looking at data, seeing how it fits into what one wants to accomplish, and then choosing a method to attain a goal.

- With a computer, it is possible to set up the "parameters" for a decision in advance. As the computer gathers data that students give it, it will try to make a logical decision about what to do next.

- A simple example would look like this. Suppose that Jane had a grandmother who lived somewhere in the United States. Suppose that Jane's mother told her that Grandmother did not live in the South or East. What is known thus far? Jane's grandmother must live somewhere in the Western or Northern halves of the United States. This helps a little. What if she also found out that her grandmother lived on the Pacific coast. This helps a great deal because now she knows that there are only a few states on the Pacific coast. Finally, what if she were also told that her grandmother lived in a state that shares a border with Mexico?

- Only one state fits all the descriptions that were listed above. Can students guess what state this is? If they figured this out they used a kind of logic to do it. This is called the "Process of Elimination." By discarding those states in which they were certain grandmother couldn't live, they narrowed the choice down to one.

- During this activity, students will be working with what are called "If-Then" statements. If something is true, then a particular actions happens, but if something is false, then a different kind of action happens.

- How does the computer know if something is true? Students tell it. The computer will take what they have told it was "true" and then compare this to see if what the "user" enters matches the true statement. Sounds kind of complicated but it really isn't.

- On the next page students will see the screen for a *Hyperstudio* project that they will be creating. This project will use "*HyperLogo*" to generate some "If-Then" statements using the "code" that goes into making this program work.

Activities Using Logic

IF-THEN STATEMENTS *(Cont.)*

On the Computer:

- This is the basic layout for the *Hyperstudio* screen. These graphic images were taken directly from the HS Art folder, but students could use graphics if they are **careful** when writing the necessary *HyperLogo* code.

- Beneath each graphic is a Text field. This is where the user will type the correct answer for each item. For example, for this graphic, the user would type the word "telephone."

- When the Check Answer button is pressed, a melody will play if the spelling of the word is correct. Each graphic works exactly the same way so that once they have written the *Hyperlogo* code for one they can copy and paste it for each of the other Check Answer buttons.

- There is something that is not seen, however. Also on this card are five invisible text fields that contain the correct spelling for each item. These invisible fields are what the Check Answer buttons are "comparing" against to see if the user typed in the correct word. This is where the "If-Then" statements come into play. If the correct word is typed, do something, as in play a melody. If an incorrect word is typed, either do something else or do nothing at all, which is the case with this program. In other words, they will not hear a melody until the correct answer is typed. What they want the computer to look for, is what the word "correct" means. Certainly an answer of "phone" for the top left picture would seem to be correct, but the computer is looking for "telephone."

- Students should first place the graphics on their card. Next, place a Text Field beneath each graphic. It is **important** to give each text field a name. A good name to label them would be **first, third, fifth, seventh,** and **ninth**. *If they skip this step, this stack will not work.*

- Next create a "Check Answer" button for each graphic. They can make all the answer buttons and text fields the same by using the copy and then paste command. *They **do not** have to name the Check Answer buttons.*

- Students now need to create the five invisible fields. These must also each have a different name. They can place these invisible text fields at the bottom right corner of the computer screen. Name each new text field second, fourth, six, eighth, and tenth.

© Teacher Created Materials, Inc.

Activities Using Logic

IF-THEN STATEMENTS *(Cont.)*

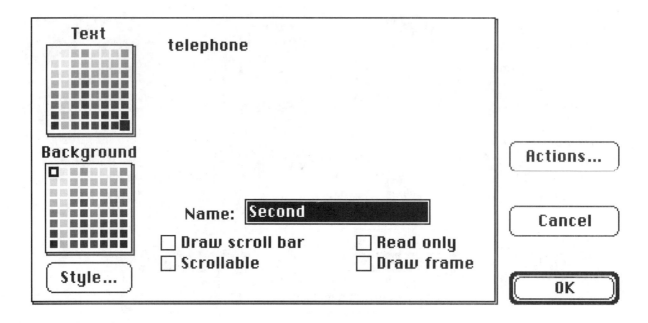

- Above they will see an example of what the "second" text field looks like as it is being set up. Notice none of the check boxes have been selected but the name has been entered as "second." In a moment they will see how to make each of these new text fields invisible to the user
- Here is how these new text fields work. Students should match up each with the first five text fields that they created. First goes with second. Third goes with fourth. Fifth goes with sixth. Seventh goes with eighth. Ninth goes with tenth.
- Once text fields second, fourth, sixth, eighth, and tenth have been created, change the browser tool back to the pointing finger.
- Click in each new text field one at a time and type the following text:

   For **second** type the word **telephone**
   For **fourth** type the word **computer**
   For **sixth** type the word **dolphin**
   For **eighth** type the word **calculator**
   For **tenth** type the word **beaver**

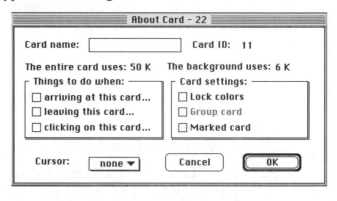

- If they need to move text fields around, remember to use the edit text object tool first to make the move. Then go back to the finger tool to type in the text for each text field.
- At this point the text fields second, fourth, sixth, eighth, and tenth, are not invisible. Students can make them invisible by adding this line of code to the "Objects-About This Card" menu item. When they make this choice, they will see a new window that looks like this sample.
- Select the option that says things to do when "Arriving at the Card" and they will see a new screen. One of the choices will be to "Use HyperLogo." When they make this choice, they will be presented with a blank screen where they can enter *HyperLogo* code.

Activities Using Logic

IF-THEN STATEMENTS (Cont.)

- On the blank screen, students should enter this *Hyperlogo* code. They should type it exactly.

    hideitem [] "second "1
    hideitem [] "fourth "1
    hideitem [] "sixth "1
    hideitem [] "eighth "1
    hideitem [] "tenth "1

- This code hides each of these text field items the minute that he or she come to this card. If his or her stack only has one card they can use the **Apple+>** shortcut to circle back to the start of the stack. Once these text fields are hidden they stay hidden until told to reappear. They actually want them to be hidden at all times, because they contain the answers the computer program will be comparing with the contents of text fields, first, third, fifth, seventh, and ninth. Remember these were the fields they created earlier in this project and they sit right under each graphic.

- The last step, they will need to do five times. This is not a problem because each bit of code will be saying the same thing, except for one change.

- They should change to the edit button tool so that they can edit the buttons named "Check Answer." Double click on the "Check Answer" button underneath the telephone. Select Actions. At the Things To Do window choose "Use *Hyperlogo*." They will now see a blank *HyperLogo* screen. Type in the following code exactly.

    make "first getfieldtext [] "first
    make "second getfieldtext [] "second
    if equalp :first :second [toot 60 60 toot 70 20 toot 60 20]

- This is what the code means. The first line says to get the text from a text field named "first." They should recall that this is the name of the text field beneath the telephone. The person using this program should enter an answer here.

- The second line says to get the text found in a text field named "second." Students should recall that this text field was made invisible and that the word "telephone" was typed into it before it was made invisible.

- The third line is the *If-Then* statement. It says that if the text in the text field named "first" matches the text in the text field named "second" play a melody. This code [toot 60 60 toot 70 70 to 60 20] actually is three notes which can be changed later if one wishes.

- If the user does not enter the correct word for the graphic image, nothing happens.

- Students can select and copy this entire bit of code and paste it into each of the other "Check Answer" buttons. The only change they will need to make is to the last **word** of lines one and two. For example here is the code for the "Check Answer" button under the computer graphic image.

    make "first getfieldtext [] "third
    make "second getfieldtext [] "fourth
    if equalp :first :second [toot 60 60 toot 70 20 toot 60 20]

- The words "first" and "second" have been replaced with "third" and "fourth."
- They will need to make this minor change on each of the other "Check Answer" buttons.
- When they are finished, they will have a card that can check for the spelling of images and determine whether the user spelled the words correctly or not. Congratulations!

Student Created Computer Programs

HYPERSTUDIO ADDING MACHINE

Procedure:

Before the Computer:

- In order to do this project, students will need *Hyperstudio 3.0* for Macintosh or Windows. This activity is for students who have worked with *Hyperstudio* before and have a general understanding of how the tools are used.

- Although this project consists of only one card, it may take several sessions to complete. Plan on 90–120 minutes on the computer.

Overview:

- This program will use some of the powerful features that are built into *HyperLogo,* which is a scripting language that comes with the *Hyperstudio* package. Although *HyperLogo* is built right into the program, most students never use it. There is already so much available within *Hyperstudio,* particularly when using NBA's or new button actions.

- The program that students are going to create, will generate a pair of numbers that they may define as any size between 1 and 33,000. The user will need to add these numbers in his head. A check button is provided to see if the entered answer is correct. Sounds easy right? Think for a moment how one might create such a program using what one knows about *Hyperstudio.* One will find that it is not as easy as it sounds.

- Once they understand how this program works and put it together, they can easily construct other programs that do multiplication, subtraction, and division. They will need to make some modifications, but these will not be difficult. A sample screen is shown below, but theirs may not look exactly like this when they are finished.

- Take a look at some of the features of this card. First, there are two graphic images. These are just there to add interest and do not do anything.

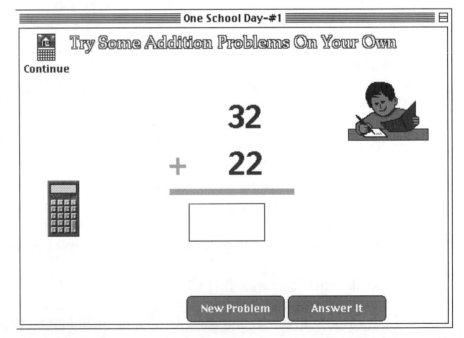

- There are four text fields. Three of them are visible at this time. The fourth one is hidden until the button "Answer It" is pressed.

- Finally there are two buttons found at the bottom of the card. One button creates new problems and the other one checks the answers entered by the user.

#2425 Integrating Technology Into the Math Curriculum   124   © Teacher Created Materials, Inc.

Student Created Computer Programs

HYPERSTUDIO ADDING MACHINE (Cont.)

On the Computer:

- Students should be ready to start their own program. Start *Hyperstudio*. If they are going to use the standard *Hyperstudio* cards they should select "New Stack" and answer yes or okay to all questions. If they wish to make their cards larger, they will need to either make that choice when they select "New Stack" or they can go to the Objects menu and select "About This Stack." This will allow them to change the size of the cards in their stacks and the number of colors. The sample project was done on a standard-sized *Hyperstudio* card.

- They should now add any two graphics that they would like to use to enhance the appearance of the finished project. These graphics will have no other use except for decoration.

- Now students should create the first two text fields. In the sample, these are the fields with the numbers 32 and 22 in them. Select Add A Text Object from the Options menu. They will need to reduce the size of the text field until they have a box that is rectangular and capable of holding four to five numbers. After they have re-sized their text fields, and clicked outside of them, they will be taken to this screen.

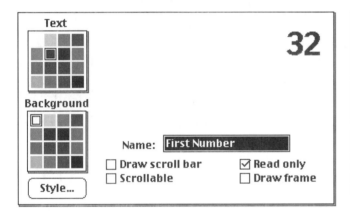 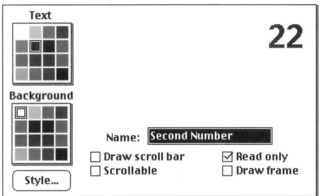

- The screen shots above show exactly how to set up these text fields. Notice some very important things about them. The one on the left is the top field on the card and is labeled First Number. The only other box that is checked off is the Read Only box. This means that the user cannot alter this box in any way. He can only see the numbers that are presented here. Students may choose any Foreground and Background colors that they wish. They can also play around with the size of the font by selecting the style button. What they want to do is to have a font that is clearly visible and fits in the text fields that they have created.

- The second image on the right shows the second text field on their cards. In every way, it is just like the one on the left, except that the name is different. The label for this text field is Second Number.

- Students should make the style, size and kind of font, the same for both text fields.

- When they are in the browse mode, the finger is selected from the tools menu, they will not even see the top two text fields. The only reason that they see them in the sample is because the "New Problem" button was pressed before the screen shot on this page was taken.

- Students should take care to get everything discussed on this page correct. If they are uncertain about what to do, ask a classroom expert.

Student Created Computer Programs

HYPERSTUDIO ADDING MACHINE (Cont.)

On the Computer: *(cont)*

- Okay let's see where students are in this project. Look again at the card layout. So far, students have created the top two text fields and have added 2 graphic images. From the screen shot of the project, they can see the items that are completed.

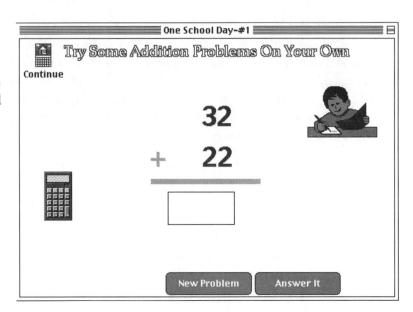

- They may have noticed that there is a graphic at the top left corner of the screen that says Continue. If they were going to build other parts to this stack that included this card they should have a button like this.

- In order to create the "+" symbol and the dividing line of the addition problem, they need to use their tool boxes. The plus sign is nothing more than a large font with the plus symbol. The solid line is created using the line tool and changing the line width by using the Options-Line Size menu. The tools that were used are on this tool bar.

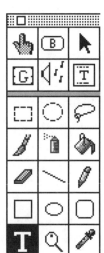

- They are now ready to create the two text fields that sit under the separation line. Remember that one of these fields will not be visible. The first new field that they will create is in the answer field under the equals line. This field is the only one with a border and is where the user types in the answer to the problem. For this sample problem, the answer would be 54.

- From the Objects menu, choose Add A Text Object. They will need to resize the text object and place it under the line. If, for any reason, they get stuck or make a mistake while re-sizing, they can always reselect the object with this tool and size it again.

- They will be presented with this screen. Make sure they choose the same options that are shown here and that they name the field Response.

- Also be sure to check the box labeled Draw frame. They can choose any colors for the Text and Background, but they will want to make the style exactly the same as the first two text fields they created earlier.

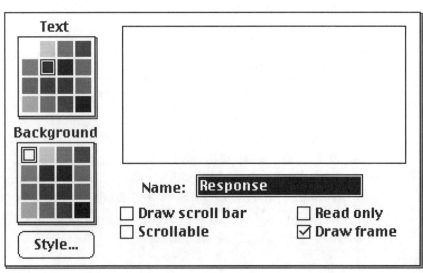

#2425 Integrating Technology Into the Math Curriculum   126   © Teacher Created Materials, Inc.

Student Created Computer Programs

HYPERSTUDIO ADDING MACHINE *(Cont.)*

On the Computer: *(cont)*

- At this point, students are ready to create the last text field. This one will be invisible except when the "Answer It" button is pressed. At this time, it will display the correct answer which can then be compared to what the user entered.
- In order to create this field, go to the Objects menu and once again select Add A Text Object. They will need to size this new text field to the same dimensions as the one above it, which is the Response text field.
- When they come to this screen make the following choices:
    a. Name this text field "Check."
    b. Check **only** the Read only box.
    c. Make the style exactly the same as the previous three text fields.
    d. Choose a text and a background color.
- *Make certain that they have named each text field correctly and that they have selected the options for each one as directed. Failure to do this will result in a program that will not work.*
- They have now finished creating all the text fields that will be needed for this project. The next step is to create the two buttons found at the bottom of the screen.
- This stack was done in 16 colors. They may see more colors for the Text and Background if they are in 256 color mode.

Creating the Buttons:

- In this project, the buttons are what make everything happen. All of the text fields are controlled by the buttons. That is why great care needs to be taken when naming the text fields.
- Create two new buttons using the Objects menu. Pick any kind of button desired, or use an icon, so long as each one is named. The button that is directly beneath the four text fields is called "New Problem." The other button is called "Answer It."
- If students want to make both buttons the same size and shape, and only change the names, they can use the Edit copy and paste functions to make duplicates.
- Move to the Places To Go-Things To Do Window. If, for some reason, students are lost, simply click on the button editor in the tool bar and then double click on a button. Select the word Actions and they will be at the Places To Go-Things To Do window.
- In the next section they are going to be using *HyperLogo* scripting to tell the text fields what they are to do. Students should be careful to copy the *HyperLogo* commands exactly. Each command will be explained later.

Student Created Computer Programs

HYPERSTUDIO ADDING MACHINE (Cont.)

HyperLogo Script:

New Problem Button:

- Students should work with the "New Problem" button first.
- In the Places To Go-Things To Do window select Use *HyperLogo*. Type the text below exactly as it appears.

    setfieldtext [] "First\ number random(100)
    setfieldtext [] "Second\ number random(100)
    setfieldtext [] "Response "
    setfieldtext [] "Check "

- The first line of these four lines of scripting, puts a random number between 1-99 into the top text field that was named "First." The second line of code puts a random number into the second text field that was named "Second." The third line clears the "Response" field of any text that might be in it from a previous problem. Finally, the fourth line of code clears the "Check" text field of any text that might be in it.

- To summarize what happens when the "New Problem" button is pressed.
    1. Puts a random number into "First" field.
    2. Puts a random number into "Second" field.
    3. Clears both the "Response" and "Check" fields of any text that might be there.

- The random numbers are what create the problem(s). The user will need to add them together and then enter an answer in the "Response" box.

Answer It Button:

- Now students should work with the "Answer It" button.
- In the Places To Go-Things To Do window select Use *HyperLogo*. Type the text below exactly as it appears.

    make "FirstNum GETFIELDTEXT [] "First\ Number
    make "SecondNum GETFIELDTEXT [] "Second\ Number
    make "theAnswer :FirstNum + :SecondNum
    setfieldtext [] "check :theAnswer

- Here is what these four lines of code mean. First of all the word **Make** is very important in *HyperLogo* because it sets up what is called a **variable**. A variable can stand for many different numbers and letters.

- If students look at the first line, they will see that the words right after **Make** are **"FirstNum."** The value for **"FirstNum** is coming from whatever is in the **"First** text field. Because this changes every time the user presses the New Problem button, it is necessary to use a variable to represent it. The words **"GETFIELDTEXT"** simply mean to get the text from a field. They will see that line two is similar. This time a new variable is used. **"SecondNum** is used to represent the text in the "Second field.

- The third line says to add up the values of "FirstNum and "SecondNum and store this total in a new variable called **"theAnswer.**

- Finally, the fourth line prints the result of "the Answer in the text field called **"check.**

Student Created Computer Programs

HYPERSTUDIO ADDING MACHINE (Cont.)

HyperLogo Script:

- It is essential that students get the "syntax" correct when working with scripting languages such as *HyperLogo*. If something is not entered correctly, they will probably see a message such as **"HyperLogo Error"** or **"To Without End."**
- Also, if the text fields are not named correctly, the program will not function correctly.
- While it may seem like a lot of work at first to use features such as *HyperLogo*, the end result is well worth it. To to create such a program as this one using *Hyperstudio* but not using *HyperLogo* would be difficult or impossible.

Assessment:

1. What was the most difficult part of this activity for you? _____
   _____
   _____

2. When you found that you were stuck, what did you do to move on? _____
   _____
   _____

3. Did any ideas for other projects using HyperLogo come to you while you were working on this one? _____
   _____
   _____

Follow Up Activity:

- If students found that they were successful with this project, they may want to try a modification.
- All the basic components are there to do a very similar program that does multiplication instead of addition. In *HyperLogo*, and many other programs, the symbol used for multiplication is the "*" which is found when you press Shift+8.
- Here are some clues as to how they could make these modifications. First, none of the text fields need to be changed in any way. Second, they will need to change the random numbers for the "New Problem" button, otherwise they will have very large multiplication problems. They will also need to change the large "+" symbol to a "x" symbol on their *Hyperstudio* card.
- The "Answer It" button needs to have *only line three* changed.
- **Good luck.**

Student Created Computer Programs

AN ASTOUNDING PROGRAM

Before the Computer:

- During this activity, students will be using *Astound* for Macintosh or Windows to create a graphical program that tests *object recognition* using a multiple choice method.

- The primary purpose is to acquaint students with the possibilities available using the program *Astound*. Although this activity will be simple in terms of layout and performance, the ideas learned can be used to generate math training presentations and programs.

- If you do not currently have a copy of *Astound,* you can download a 30-day trial copy that is functional (including save capabilities) from http://www.astound.com. Versions are available for Windows 3.1, Windows NT, Windows 95, and Macintosh.

- Before they actually start doing this project, here is an overview of what *Astound* can do. Many people use *Astound* to do presentations of all sorts. Because it comes with a large number of built in "templates," putting together a presentation is a snap. Many "templates" or "master pages" come with animations, fade in and fade out effects, and other features that are similar to *Hyperstudio*. In addition, you can work with text, graphics, sound, animation, video, and much more. In this respect, *Astound* is also similar to *Hyperstudio*.

- *Astound* can take the input of a user and do something with it. If, for example, the user presses a button, an action can be assigned to this button.

- During this activity, students will be creating a number of screens that have a single graphic and three choices below the graphic. These are actually 3 buttons with different words written on them.

- This screen is an example of what the *Astound* window looks like running on Windows 95.

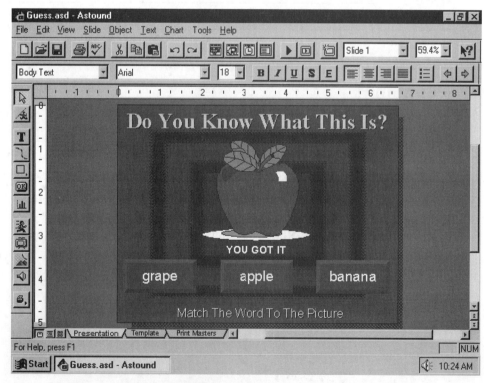

- This is a screen shot of the actual project. The area where students work on slides is found here. This is what Slide 1 will look like, except that the graphic they choose may be different. The tools will be explained as they work with the project. Most of the tools they will be using will be found on this tool bar.

Student Created Computer Programs

AN ASTOUNDING PROGRAM (Cont.)

- Now take a look at Slides 1, 2, and 3. As the program is running, the slides become full screen size and they wait for the user to input an answer. Study each of these three slides and details about them will be discussed.
- Each slide in the project has some common elements. First, the background pattern is always the same. Second, each slide has the same title, "Do You Know What This Is?" Third, there is a single graphic and then three "buttons" beneath the graphic that contain some words.
- Like a multiple choice quiz, only one of the answers is correct. When a student presses the correct answer, a message appears right beneath the graphic. The message can not be seen at the moment because no buttons have been pressed.
- By using copy and paste techniques that they have already used in other projects, students will be able to speed up the time that it takes to create their slides.

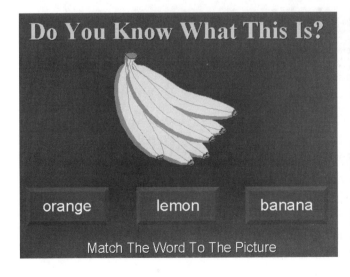

Student Created Computer Programs

AN ASTOUNDING PROGRAM (Cont.)

- When students first launch *Astound*, the opening screen presents the user with a number of choices. Select the one that says "Create From A Blank Template." Students will see a blank screen with two "text place holders." These place holders are already programmed to perform certain functions.

- In the large text place holder that says "Type Presentation Title Here," they are going to enter the text "Do You Know What This Is?" This is done by double clicking in the text place holder and then typing the text. They can easily adjust the text size and they can also stretch out their place holders so that they cover most of the screen. Just grab any corner to stretch, and click in the middle of the place holder to move it. The type of font and the size are listed right above the work window.

- Now, double click on the smaller text place holder found at the bottom of the window. Change the text here to read "Match The Word To The Picture."

- The background color is called a gradient fill. In order to get this effect use the "Slide" menu and then select Background. There are many choices here, and students can experiment with a background color.

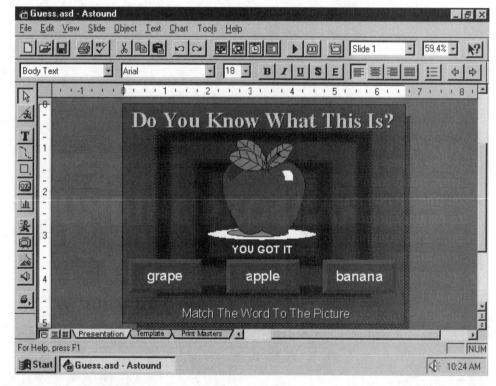

- Text color can also be modified by using the Quickset tool. They will need to first click once on the text fields and then choose new fill colors.

- The last thing they will want to do is to add their buttons and then copy their slides twice. For now the buttons will be non-functional. Later they can assign actions to them. In order to create buttons, and add some text, they will need to use two tools. Both of these are found along the left side of the computer screen. The button tool looks like a rounded rectangle with the letters OK in the middle. The text tool is a capital T.

- Click on the button tool and go over to the work area. "Drag" out a rectangular shape and then click outside of the shape. They should see gray buttons that look raised. Place these buttons near the bottom of the screen as in the example. They are going to use the copy and paste commands found in the Edit menu. Click on one of the buttons to select it and then choose Copy. Now choose Paste. Reposition the new buttons and then choose Paste again. Line up all three buttons as in the sample.

Student Created Computer Programs

AN ASTOUNDING PROGRAM (Cont.)

- At this point, students should have one side that contains a text field at the bottom and a text field at the top. They should have also used a gradient or some other kind of fill pattern on this slide so that it is colored. At the bottom of the screen, there should be three buttons that are lined up horizontally. If they have all of this done, then they are ready to proceed.

- Before going any farther, they are going to duplicate their slides two more times. Here is what they need to do. From the Slide menu choose the word Duplicate. When they do this, the number will change to Slide 2.

- Duplicate again and they will be at Slide 3. Return back to Slide 1 by using the same pull down menu.

- They can choose what kind of graphic to place on each of their slides. For the purpose of this lesson, they should choose apples.

- Using the tool bar along the left side of the computer window, select the graphics tool. Depending on what system they are using, Macintosh or Windows, the types of images they may have available will be different. They can use any clip art library available or a CD-ROM.

- They will be taken to a new window that asks if they wish to select this graphic for placement on the slide. Choose "yes" and then position the graphic where desired.

- Graphics can also be resized in the same way that text fields were earlier. They can make them smaller or bigger as the need arises.

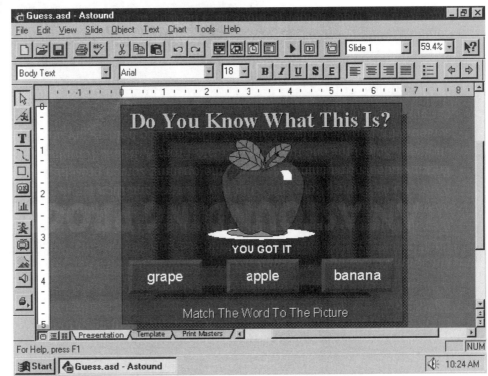

- After each student has finished placing a graphic of his/her choice on Slide 1, do the same thing for Slides 2 and 3. Make sure that students know how to spell the words that are associated with each graphic.

- In the next section, they are going to finish this project by adding words to each button and then telling the button what to do when the user selects it.

- For now they can test their programs even though they are not finished. When they run these tests each slide will show for about 15 seconds and then bring on the next slide. Click on the button located below and to the right of the word "Help." It looks like a triangle or a "Play" button found on most VCR's. Afterwards, they will have to press ESC to return to the work window.

Student Created Computer Programs

AN ASTOUNDING PROGRAM (Cont.)

- In review, students should each have three slides, each containing a graphic, two text fields, and three buttons at the bottom of the screen. If they have tested their programs, they should have seen each of their slides displayed after a 15 second wait between slides.

- Students should move back to Slide 1 and add some text to each of these buttons. Click on the "T" tool once and then click on the buttons that they want to name. Now, they should see blinking cursors in the buttons and they can type a word. Each will need to decide in advance what three words he or she is going to use. Only one of the three is going to be the correct answer. In the sample, apple is the middle word while the other two are incorrect.

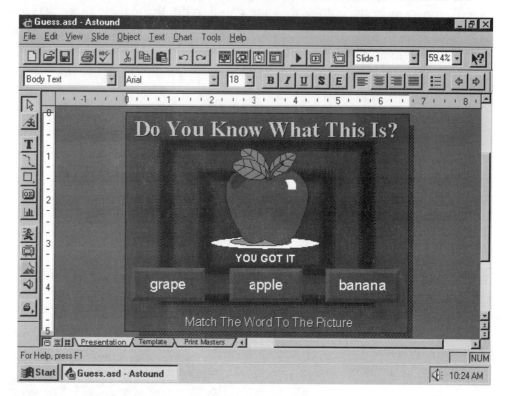

- After they have given names to all three buttons on Slide 1, they will be ready to edit the actions that these buttons will perform. For those buttons that are not the correct answer, double click on the button. This will bring up a dialog box that contains many functions. It is called the "Object Interaction" window.

- Under "Triggers Events," check the box next to the word mouse. Under the "Sound" tab, select the choice that says "Plays System Beep." This means that when the user selects the incorrect answer the system will beep (for Windows Users).

- Do the same thing for the other incorrect answer button.

- For the correct answer button, double click on it and students will be at the "Object Interaction" window.

- Select "Mouse Triggers Events."

- Under sound select "None."

- Under the "Flow" tab select "Relative Slide." Make the relative slide the "Next Slide" by choosing it from the pull down menu.

- They will need to repeat these steps for all of their slides. Additional slides can be added at a later time.

Student Created Computer Programs

AN ASTOUNDING PROGRAM *(Cont.)*

- Students may want to experiment with the timing of their slides. Right now each slide will display for 15 seconds. When the user selects a wrong answer he/she will hear a beep. If the user selects a correct answer there will be no beep but at the end of 15 seconds the next slide will appear. If students would like to make the time interval different for all of the slides, they can make this change by using the "Slide" menu. Select "Properties" and simply enter a new number. Each slide can be individually changed in this manner.

- A student could also add a visual message like the one found in the sample. Create a new text field and make its contents say "YOU GOT IT." Using the "Object" properties menu. give this text field a name such as "Got It." As a student is editing the correct answer button, the "Object Interaction" window will have a tab called "Action." Select the choice that says "Control Object." To the right of this choice are two pull down menus that are called "Action" and "Object." Set the "Action" to start time line and set the "Object" to "Got It (text)" or whatever they named it.

- Now when the correct answer is chosen, there is a visual message box and then the slide moves forward.

- There are many other features of *Astound* that are beyond the scope of this book. Hopefully, teachers and students will be curious and download a demonstration copy of *Astound* from their web site at **http://www.astound.com**.

Assessment and Follow-up:

As with many other projects in this book, the true test of whether you understood what was to be done is whether or not your program functioned as described.

Beyond this objective is the hope that you can see the potential of using this program, or others like it, as a training/teaching tool for yourself and others.

- List some of the ways that you can see a program like *Astound* being useful to you both in the classroom and at home.

    1. _____
       _____

    2. _____
       _____

    3. _____
       _____

Resources and Additional References

PROJECT CHECKLIST FOR TEACHERS

Student	P-13	P-15	P-18	P-20	P-22	P-25	P-28	P-31	P-34	P-36	P-41	P-44	P-48	P-50	P-52	P-53	P-55	P-57	P-60	P-62

Resources and Additional References

PROJECT CHECKLIST FOR TEACHERS

Student	P-	P-	P-	P-	P-	P-	P-	P-	P-	P-	P-	P-	P-	P-	P-	P-	P-	P-	P-

Resources and Additional References

PROJECT CHECKLIST FOR TEACHERS

Student	P-	P-	P-	P-	P-	P-	P-	P-	P-	P-	P-

SCHEDULING

COMPUTER SIGN-UP SHEET

WEEK OF _____

Students: Be sure to fill in your name, project, reason you need to work on it, and the time needed to finish it. If any of these columns are left blank, your request will not be considered. Leave the Scheduled Time column blank so you can be assigned a time.

Student Name	Project	Reason	Time Needed	Scheduled Time

Resources and Additional References

Resources and Additional References

PLANNING

STATUS OF THE CLASS MANAGEMENT SHEET

Project: _____   **Project Dates:** _____

Student Name												

Resources and Additional References

PROJECT ANSWERS

- This activity involves following the directions found on page 114. Below, students will see how the answers to the three problems are supposed to look. Although these answers are not in color, theirs should be.

Problem #1:

Problem #2:

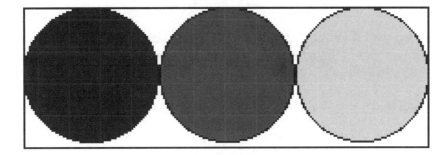

Problem #3:

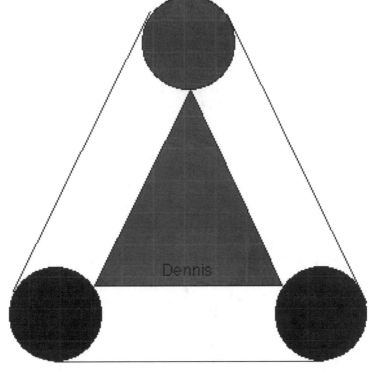

© Teacher Created Materials, Inc. 141 #2425 Integrating Technology Into the Math Curriculum

Resources and Additional References

SOFTWARE BIBLIOGRAPHY

Many software titles were mentioned at the beginning of this book. A few of them are in the "must buy" category. These are perhaps the easiest to use, the most affordable, and are mentioned often throughout this book.

Claris Corporation: 5201 Patrick Henry Drive, Santa Clara, CA 95052 1-800-544-8554

ClarisWorks 4.0 is a multifunction software package that combines word processing, drawing, painting, spreadsheets, database, and communications capabilities. There are a multitude of ways to use this program both in the classroom and office.

ClarisWorks for Kids is a brand new program that was just released as this book was being written. It has had excellent reviews and beta testing results were good. This program would be reccomended for elementary students.

Claris Home Page is a web page development program that was not specifically mentioned in this book but would be very useful for Internet purposes.

Sunburst Corporation: 101 Castleton Street, P.O. Box 100, Pleasantville, NY 10570-0100 1-800-321-7511

Type To Learn is an award winning program that teaches basic keyboarding. This is an essential skill when working with computers in any manner. The author believes that this is the best program on the market and comes at a very reasonable price tag. (site license)

Broderbund Software: P.O. Box 6125 Novato, CA 94948-6125 1-800-474-8840

Kid Pix 2 allows students to create pictures, slide shows, animations, and also record sound. Many activities in this book can be done using *Kid Pix 2*.

Kid Pix Studio is an enhanced version of *Kid Pix 2* and allows for neat special effects, movies, and a host of other multimedia features that are similar in some ways to *Hyperstudio*.

Roger Wagner Publishing: 1050 Pioneer Way, Suite P., El Cajon, CA 92020 1-800-421-6526

Hyperstudio 3.0 for Macintosh and Windows is one of the most versatile multimedia programs on the market today. It has a very short learning curve and is user friendly. Children will get results quickly. Although simple to use, *Hyperstudio* can also be a very powerful program that can create professional looking presentations.

Astound Incorporated: P.O. Box 59, Santa Clara, CA 95054 1-800-982-9888

Astound 4.0 for Windows and *Astound* 3.0 for Macintosh is a terrific multimedia publishing program with many of the same kinds of features found in *Hyperstudio*. Geared a little more toward business presentation, is can easily be used for nearly all of the projects that use *Hyperstudio* in this book.

The Learning Company: 6493 Kaiser Drive, Freemont, CA 94555. 1-800-852-2255

The Writing Center is an excellent word processing program for elementary students.

Resources and Additional References

OTHER RESOURCES

Where to buy Software:

- Most school districts have limitations on what vendors they can use. Because purchase orders often take a long time to be processed, it is usually difficult to shop through those outlets that provide the best price. The sale is over before the money is avaiable. For both the Mac and the PC, three excellent choices for purchasing software, as an indivdual or school district are from:

  The Mac/PC Zone- 1-800-248-0800
  Mac/PC Warehouse- 1-800-696-1727
  Mac/PC Mall- 1-800-222-2808

- Call any of the 800 numbers listed above and they will send you a catalog. Each of these companies consistantly has the best price on computer products and often will match any price that you find lower. In addition shipping is very reasonable.

Other Recommended Sources:

Edmark Corp.
P.O. Box 3218
Redmond, WA 98073-3218
(800) 426-0856

Grolier Electronic Publishing
Sherman Turnpike
Danbury, CT 06816
(800) 356-5590

MECC
6160 Summit Drive North
Minneapolis, MN 55430-4003
(800) 685-6322

Microsoft Corporation
One Microsoft Way
Redmond, WA 98052
(800) 426-9400

National Geographic Educational Software
P.O. Box 98018
Washington, D.C. 20090-8018
(800) 368-2728

Troll Associates
100 Corporate Drive
Mahwah, NJ 07498-0025
(800) 526-5289

Resources and Additional References

BIBLIOGRAPHY

Balka, Don. *Attribute Logic Block Activities.* Ideal School Supply Company, 1985.

Basile, Leonard J.. *Math Maneuvers.* Dale Seymour Publications, 1987.

Burns, Marilyn and Bonnie Tank. *A Collection of Math Lessons.* The Math Solution Publications, 1988.

Kaye, Peggy. *Games For Math.* Pantheon Books, 1987.

Lynn, Bill and Mike Westerfield. *HyperLogo-A Scripting Language for Hyperstudio.* Roger Wagner Publishing, 1995.

McBride, Karen. *Map Skills.* Bowmar/Noble Publishers Inc., 1977.

Parker, Ruth E. *Mathematical Power-Lessons From A Classroom.* Heinemann, 1992.

Russell, Susan Jo and Rebecca B. Corwin. *Used Numbers-Real Data In The Classroom.* Dale Seymour Publications, 1989.

Online Services

America Online, (800) 827-6364

Earthlink, (800) 395-8425

Internet Interfaces

Netscape Navigator 2.0 and 3.0

Microsoft Internet Explorer 2.0, 3.0